T0290356

Citizen-Soldiers and Manly Warriors

Military Service and Gender in the Civic Republic Tradition

R. Claire Snyder

ROWMAN & LITTLEFIELD PUBLISHERS, INC.
Lanham • Boulder • New York • Oxford

ROWMAN & LITTLEFIELD PUBLISHERS, INC.

Published in the United States of America
by Rowman & Littlefield Publishers, Inc.
4720 Boston Way, Lanham, Maryland 20706
http://www.rowmanlittlefield.com

12 Hid's Copse Road
Cumnor Hill, Oxford OX2 9JJ, England

British Library Cataloguing in Publication Information Available

Library of Congress Cataloging-in-Publication Data

Snyder, R. Claire, 1965–
    Citizen-soldiers and manly warriors : military service and gender
in the civic republican tradition / R. Claire Snyder.
        p.   cm.
    Includes bibliographical references and index.
    ISBN 0-8476-9443-7 (alk. paper). — ISBN 0-8476-9444-5 (pbk. :
alk. paper)
    1. Military service, voluntary.   2. Citizenship.   3. Masculinity.
I. Title.
    UB320.S64   1999
    306.2′7—dc21   99–23275
                                                    CIP

Printed in the United States of America

♾ ™ The paper used in this publication meets the minimum requirements of American
National Standard for Information Sciences—Permanence of Paper for Printed Library
Materials, ANSI/NISO Z39.48—1992.

For my family:
Lee Daniel and Anne G. Snyder, Tim and Lisa Snyder, and Allison Turkel

Contents

Acknowledgments

In finishing this project, which truly marks the beginning of my career as a political theorist, I have been reflecting a lot on the past decade and on all the debts of gratitude I owe to the many people who helped and supported me along the way. This book began as a doctoral dissertation at Rutgers University and so to the political science department there I give thanks. I particularly want to thank the faculties in political theory and in women and politics for providing me with an excellent education that has enriched my life in more ways than I can recount.

In addition, I am intellectually and otherwise indebted to the Walt Whitman Center for the Culture and Politics of Democracy and to the Charles F. Kettering Foundation for their generous support over the years. I would also like to express my appreciation to my colleagues in the department of political science at Illinois State University and particularly to my visionary chair, Jamal Nassar, who has provided me with unflagging support during my hectic first year of full-time teaching.

My thinking on the topic of the Citizen-Soldier has benefited greatly from presenting parts of this work at the Conference on the Public Sphere at the Institute for Research on Women, the Seminar on War and Peace at the Rutgers Center for Historical Analysis, various meetings at the Charles F. Kettering Foundation, the Twelfth Annual Symposium on the History of the Navy at the United State Naval Academy, and annual meetings of the Northeast Political Science Association.

I am grateful for all the helpful criticism and guidance I received from my stellar dissertation committee during the initial stages of this project. First of all, I would like to thank Charles Moskos for serving as my "outside reader" and for all the comments he provided on the manuscript. Next, I want to extend a huge thank you to Stephen Bronner, whose intellectual analysis and political vision originally inspired this project. My deep and heartfelt gratitude goes to Linda Zerilli, who more than anyone has made me see the world in a completely different way. Over the past decade, Linda has consistently been both challenging and supportive of my work and of my development as a feminist political theorist, and for that I am truly grateful. And finally, I want to thank Benjamin R. Barber, who encouraged me to pursue this project from the beginning, who successfully negotiated the delicate balance between directing my doctoral research and allowing me room to develop my own voice, and who has provided me with much-appreciated advice and support over these many years.

I would also like to thank my editor at Rowman & Littlefield, Stephen Wrinn, for his enthusiastic commitment to this work and for making this book possible. In addition, I would like to thank Stephen Driver, Julie Kirsch, and Mary Carpenter for all of their help with the manuscript and the publication process in general. And a final thanks goes to my personal research assistant, Abdalla Khair, for all of his help during the final stages of this project.

This book truly would not be what it is today without the help and advice of my friend, colleague, and comrade Manfred Steger, who spent an incredible amount of time poring over the entire manuscript and providing invaluable intellectual insights as well as logistical advice, support, and friendship over the past year.

But most of all, my largest debt of gratitude goes to my family. I thank my parents, Lee Daniel and Anne G. Snyder, for providing the familial milieu in which my intellectual abilities were nurtured from an early age. I thank them, my brother, Tim, and his wife, Lisa, for believing in me, for supporting me, and for loving me all of these years. And finally, to Allison Turkel I owe everything—for her love, her generosity, her support, her encouragement, her sense of humor, and her intellectual insight over the past three and a half years. I dedicate this book to my family, for they have given me gifts I treasure dearly and can never fully repay.

R. Claire Snyder
Normal, Illinois

Chapter 1

Introduction

The *New York Times Magazine* recently ran a cover story that asked the following questions: Does "being a good soldier [depend] on being an aggressive male?" "Is there something uniquely male about the warrior? Can the warrior survive the feminization of the military, or are we sacrificing military effectiveness on the altar of political correctness?" These are challenging questions and ones that come up again and again, as America's traditionally masculine military attempts to integrate women into its ranks. Yet despite the proliferation of commentary on gender integration and a multiplicity of related problems, mass media treatments of the issues remain for the most part either journalistically superficial or highly ideological and thus never seem to move beyond the apparently irreconcilable conflict between gender equality and military effectiveness.

This book attempts to clarify, deepen, and advance contemporary debates about gender and military service by reintroducing the concept of *democratic citizenship*—a concept that is surprisingly absent from the current discourse. Although the particulars surrounding gender integration of today's armed forces are undoubtedly unique in world history, late-twentieth-century Americans are hardly the first to think about the relationship between gender and military service—and the relationship of both to citizenship. Indeed, discussions of these connections are as old as political theory itself.

Dating all the way back to ancient Greece, the civic republican tradition directly addresses the interconnections between gender, military service, and citizenship through its central ideal of the manly Citizen-Soldier. Standing at the very center of the civic republican tradition, this figure embodies the twin practices of civic republican citizenship: military service and civic participation. Citizen-soldiers serve in the military in order to protect their ability to govern themselves for the common good, and they participate in the process of deciding when to engage in war. Both halves of the ideal are equally important. Normative rather than empirical, the Citizen-Soldier ideal necessarily entails a commitment to a set of universalizable political principles, including liberty,

1

equality, camaraderie, the rule of law, the common good, civic virtue, and participatory citizenship. In other words, the Citizen-Soldier ideal cannot be reduced to universal military service; it must also include a commitment to substantive participation in the processes of self-government.

Setting aside for the moment the historically masculine character of the Citizen-Soldier, we can see that this tradition provides important insights for us today because civic republicanism remains the only political theoretical discourse that recognizes the military as a central problem for democratic society. Hierarchical institutions of coercion, such as the police and the military, always pose a problem for a democratic society in which freedom and equality are fundamental values. Nevertheless, democratic societies require the existence of such institutions to protect themselves from those who would undermine the fragile ideals of liberty, equality, and the rule of law. Or to put it in terms of the tradition itself, a democracy must vigilantly guard itself against the threats posed by internal and external enemies. Because the Citizen-Soldier tradition directly addresses the potential contradiction between civic and martial imperatives, remembering this tradition should add a new dimension to current debates about the military, whereas overlooking this tradition allows us to refrain from taking the military seriously as a part of democratic society.

This book uses the mythos of the Citizen-Soldier as a prism through which to view a variety of issues, including not only the role of the military in democratic society but also the nature of citizenship and the constitution of gender. I argue that within the civic republican tradition, the Citizen-Soldier ideal fuses military service and participatory citizenship as well as citizenship and masculinity. More specifically, the Citizen-Soldier functions as a prescriptive ideal that calls for male individuals to engage in the civic and martial practices that constitute them as masculine republican citizens. At the same time, the masculine character of the ideal undermines the participation of female individuals in civic and martial practices because these practices constitute not just citizenship but also masculinity. Consequently, because of the practical and theoretical interconnections among citizenship, military service, and gender within civic republicanism, all attempts to resuscitate this tradition as a remedy to the ills of liberal democracy must address the following question: *What happens in a tradition that links citizenship with soldiering when women become citizens?*

My book reviews and reimagines traditional republican citizenship in light of contemporary feminist theory; it also intervenes in and refocuses traditional feminist debates over citizenship. Until now the question for feminists has been how to expand the conventional understanding of citizenship to include the activities in which women have traditionally engaged and/or how to demonstrate that the historical traditions of citizenship are inherently masculine and so can never be extended to women. In these debates the categories of "men," "women," and "citizens" are generally treated as prepolitical. That is, we assume there are "men," "women," and "citizens" who then enter politics. Even social constructionists like Wendy Brown make this assumption. For example, by emphasizing that "manhood constructs politics," she argues that prepolitical, cultural understandings of "manhood" directly affect the shaping of politics because men make politics.[1]

Citizen-Soldiers and Manly Warriors seeks to sever the connection between manhood and politics in the hope of reconstructing citizenship so that it meets the standards of true universality. Relying on my own reinterpretation of Judith Butler's notion of "performativity," this study rereads civic republican theory and practice, using the idea that civic identity—like gender identity—is performatively constructed.[2] My interpretation seeks to demonstrate that civic republicanism has implicit within it a performative understanding of identity: Citizenship does not exist prior to engagement in politics. To the contrary, within the Citizen-Soldier tradition, individuals *become* republican citizens *only* as they engage together in civic and martial practices. Only through such participation do they come to manifest the feelings of patriotism, fraternity, and civic virtue that form the necessary foundation for the possibility of government aimed at the common good. Moreover, individuals never finally *become* citizens in the sense that they will always think and act in terms of the common good. It is only within the context of republican institutions, which require them to behave as citizens, that they will act and think as citizens. Hence, the process of becoming a citizen is never finished; citizens must be constantly re-produced through the repetition of particular civic practices.

In its ideal form, civic republican citizenship—what I call the *citizenship of civic practices*—contrasts with two other conceptions of citizenship: *ius solis* (citizenship tied to place of birth) and *ius sanguinis* (citizenship based on bloodlines).[3] *Ius solis*—characteristic of liberal democracies—defines citizens as any group of individuals living in a particular bounded territory, whereas *ius sanguinis*—or what could be called a *citizenship of blood*—restricts citizenship to members of a particular ascribed group. In contrast to these two conceptions, a *citizenship of civic practices* requires engagement in civic and martial practices. It is not enough to live within the borders of a nation-state, nor is it enough to have a particular ethnic identity. Rather, one's identity as a citizen-soldier requires repeated engagement in civic and martial practices.

Additionally, as we would expect, the performative understanding of identity implicit to civic republicanism applies to gender as well as to civic identity. The same practices constitutive of citizen-soldiers also construct masculinity—what it means to be a man within a civic republican framework. But while participation in civic and martial practices transforms male individuals into masculine citizen-soldiers, the exclusion of female individuals from these same practices contributes to the dominant construction of "femininity." Women traditionally engage in the practices constitutive of "republican mothers" rather than citizen-soldiers.

Understanding civic and gender identity as performatively constructed creates important opportunities for democratic and feminist theorists interested in creating a more just society. That is, if gender is performatively constructed, rather than rooted in nature, this means that gender identity is malleable rather than fixed. Because gender is constructed, it can also be reconstructed in a way that does not advantage one particular gender over another. That is, if masculine citizens are traditionally constituted through engagement in civic practices, and feminine subjects are traditionally constituted through the exclusion from civic practices, then what would be the social and political consequences of "women" engaging in the civic practices constitutive of masculine republican citizen-soldiers?

In her seminal study *Gender Trouble,* Judith Butler seems to suggest the possibility of actively changing cultural constructions of gender: "When the constructed status of gender is theorized as radically independent of sex, gender itself becomes a free-floating artifice, with the consequence that *man* and *masculine* might just as easily signify a female body as a male one, and *woman* and *feminine* a male body as easily as a female one."[4] Consequently, "through the mobilization, subversive confusion, and proliferation of precisely those constitutive categories that seek to keep gender in its place by posturing as the foundational illusions of identity" culturally mandated configurations of gender can be reworked and "gender trouble" can be made.[5] Put differently, if "men" and "women" are constantly becoming gendered as they participate in behaviors required by cultural norms of masculinity and femininity, then their transgressive engagement in counterhegemonic gender behavior should alter the cultural construction of gender and the sexism it generates. That is, the *subversive transgender performances* of "women" acting out "male" scripts could work to highlight the artificiality of normative constructions of gender and so undermine the sexism such constructions generate.

My project asks us to begin imagining the possibility of radically reconstructing not only republican citizenship but also gender itself—at least as it relates to the civic realm. Interestingly, while Butler seems to hint at this possibility in *Gender Trouble,* she quickly retreats from such constructive possibilities in her subsequent work *Bodies that Matter.* Attempting to clarify "questions raised by the notion of gender performativity introduced in *Gender Trouble,*" Butler explains that

> if I were to argue that genders are performative, that could mean that I thought that one woke in the morning, perused the closet or some more open space for the gender of choice, donned that gender for the day, and then restored the garment to its place at night. Such a willful and instrumental subject, one who decides *on* its gender, is clearly not its gender from the start and fails to realize that its existence is already decided *by* gender. Certainly, such a theory would restore a figure of a choosing subject—humanist—at the center of a project whose emphasis on construction seems to be quite opposed to such a notion.[6]

Though agreeing with Butler that choosing a gender is not as simple and voluntaristic as choosing to don a skirt or fatigues, I have many fewer reservations about placing the notion of performativity within a humanistic political framework. Indeed, I argue for the importance of restoring to debates about the construction of gender a stronger sense of agency than is possible within Butler's postmodern paradigm.[7]

Despite her theoretical agenda, Butler paradoxically insists that performativity actually entails a lack of freedom: "Performativity is neither free play nor theatrical self-presentation; nor can it be simply equated with performance."[8] Performance, she tells us, does not occur as a matter of choice but only through the "forced reiteration of norms." Moreover, not only are gender performances constrained, but in fact constraint operates as "the very condition of performativity." Indeed, Butler insists that "repetition is not performed *by* a subject"; there is no "doer behind the deed."[9] All one can hope for is a micropolitics of parody—a strategy, she stresses, that ultimately may fail to subvert dominant gender norms.[10]

Thus, Butler's postmodern framework ironically truncates the very possibility of a radical reconstruction of gender that her earlier work seems to suggest, because her rejection of a "doer behind the deed" eliminates the possibility of meaningful agency. Seyla Benhabib emphasizes this point in her critique of Butler's theoretical paradigm. Rejecting the "strong" version of postmodernism advocated by Butler, Benhabib accepts the insights of postmodern theory only in their "weak" formulations. That is, she maintains that we can accept the "situatedness" of the subject within "the context of various social, linguistic, and discursive practices" while also holding on to "the desirability and theoretical necessity of articulating a more adequate, less deluded, and less mystified vision of subjectivity."[11] In other words, we can accept the performative construction of identity without completely annihilating the concept of the subject that forms the necessary prerequisite for meaningful agency. Thus, the political goal, as Benhabib puts it, is to "stop the performance for a while, . . . pull the curtain down, and let it rise only if one can have a say in the production of the play itself."[12] Conversely, Butler exposes the performative construction of identity only to undercut the radical implications of her own argument by denying the possibility of a humanist subject capable of rewriting her own scripts.

Furthermore, Butler's deconstructionist move also destroys the normative basis upon which to argue for the justness of making such changes. And without a normative theory of social justice and human dignity, no basis exists for distinguishing between subversive acts that further liberty, equality, and the rule of law and those that undermine those fragile ideals. To be sure, no one questions Butler's allegiance to politically progressive politics; however, as Martha Nussbaum rightly points out, "Butler cannot explain in any purely structural or procedural way why the subversion of gender norms is a social good while the subversion of justice norms is a social bad." Expanding on this line of argument, Nussbaum notes:

> There is a void, then, at the heart of Butler's notion of politics. This void can look liberating, because the reader fills it implicitly with a normative theory of human equality or dignity. But let there be no mistake: for Butler, as for Foucault, subversion is subversion, and it can in principle go in any direction. Indeed, Butler's naively empty politics is especially dangerous for the very causes she holds dear. For every friend of Butler, eager to engage in subversive performances that proclaim the repressiveness of heterosexual gender norms, there are dozens of others who would like to engage in subversive performances that flout the norms of tax compliance, of non-discrimination, of decent treatment of one's fellow students. To such people we should say, you cannot simply resist as you please, for there are norms of fairness, decency, and dignity that entail that this is bad behavior. But then we have to articulate those norms—and this is what Butler refuses to do.[13]

In this passage, Nussbaum cogently articulates the major political problem with a full-scale acceptance of the strong version of postmodernism. That is to say, postmodernism does a good job of exposing the "constructedness" of the subject by political, historical, and ideological forces beyond its control. However, the postmodern rejection of humanist ideals as simply another mask for power actually eliminates the moral basis for opposing those particular forces that we consider unjust.

Heeding Benhabib's and Nussbaum's critique, *Citizen-Soldiers and Manly Warriors* takes up the notion of performativity but places it within a normative framework of social justice provided by the democratic tradition of civic republicanism. Remembering the normative parameters of this tradition is critically important, not only for the reasons just outlined, but also because of the risks inherent in the tradition itself. That is, the practices characteristic of the Citizen-Soldier tradition do indeed generate a series of virtues—patriotism, fraternity, and civic virtue—that are the essential prerequisites for government aimed at the common good. However, these same practices can just as easily produce a set of corresponding vices—including chauvinism, exclusion, and conformity. Consequently, democratic theorists who seek to reap the benefits of the Citizen-Soldier tradition must also figure out how to minimize its corresponding vices; the oppositional tendencies within this tradition cannot simply be wished away. This book attempts to solve this perennial puzzle while also reinterpreting the tradition in light of contemporary commitments to gender equality. In so doing, I foreground the normative principles that characterize the political theory of the Citizen-Soldier—its commitment to liberty, equality, camaraderie, the rule of law, the common good, civic virtue, and participatory citizenship—and argue that these elements provide normative guidelines by which to judge particular practices and particular political outcomes.

Using a performativity lens to reread the normative tradition of the Citizen-

Soldier raises the possibility of reconfiguring republican citizenship so that it meets the standards of true universality. I am interested in reconstructing republican citizenship so that *all* individuals, regardless of gender, can be included as republican citizens. In other words, I want to affirm the humanist ideals of civic republicanism—such as freedom, equality, meaningful agency, autonomy, and universalizable principles—while simultaneously exposing the shortcomings of actually existing practices through the limited deployment of some useful postmodernist conceptual tools. In other words, while I reject the notion of the "unencumbered self," I do not reject the concept of the self altogether. Instead, I demonstrate the ways in which identity is produced by the fusion of ideals and practices, and then I investigate the possibility that a contemporary reinterpretation of traditional democratic ideals combined with a new set of civic practices might yield a new form of civic identity, one that includes "women" as well as "men."

At the same time, however, I want to stress that the reconfiguration of republican citizenship to include "women" must also entail the reconfiguration of gender—of "women" and "men" as social categories. That is to say, historically "masculine" categories, such as "citizen" and "soldier," cannot simply be expanded to include "women" but otherwise remain unaltered. For example, the *armed masculinity* of contemporary soldiers is a precarious cultural construct constituted in hostile opposition to "femininity," whether located in "women" or within "men" themselves. Simply inserting "women" into a misogynistic warrior culture does not eliminate the conflation of soldiering with "masculinity," but rather produces sexual harassment and rape, as evidenced by the broad array of recent scandals within the American military. Because of traditional dichotomous constructions of gender, female individuals are viewed not as "soldiers" but as "women."

Thus, in order to render our supposedly democratic military more inclusive, we must begin to unsettle and then reconfigure our understandings of gender. More specifically, we need to move away from the idea that male individuals are "masculine" and female individuals are "feminine" and begin to see "masculinity" and "femininity" as particular sets of practices in which all individuals engage at various times. Reviving the Citizen-Soldier tradition cannot simply consist of reattaching *armed masculinity* to a resuscitated republican citizenship. Instead, we must reform the type of masculinity constructed within our military so that it does not require the hostile denigration of "femininity."

But why even try to resuscitate the Citizen-Soldier tradition rather than just rejecting it wholesale? For one, this tradition provides us with a democratic legacy and a set of democratic principles with which we can strive to reform the military and purge it of antidemocratic practices such as misogyny and homophobia—neither of which is necessarily essential to military effectiveness. Instead of simply ignoring the military realm, the Citizen-Soldier tradition places military institutions at the center of the democratic project.

Second, to reject the Citizen-Soldier tradition in its entirety would be to give up on the American tradition that anchors our calls for a more participatory form of citizenship—civic republicanism. The ideal of substantive popular sovereignty comes directly out of the civic republican tradition, which has at its center the Citizen-Soldier. Because the grand foundationalist fictions of modernity have been challenged by postmodern thinkers, it has become harder for us to anchor normative claims in unshakable "foundations." For this reason, it is useful to work within an already existing tradition. Despite its many risks, the Citizen-Soldier tradition contains valuable democratic elements whose revival could greatly improve citizenship and democracy in America.

I want to rework the Citizen-Soldier tradition of civic republicanism, because it presents us with a tradition of participatory citizenship and a commitment to universalizable principles. Right now in America, the idea that we should have government for the common good has come under sustained attack. Our society often lacks the essential prerequisites for government aimed at the common good—patriotism, camaraderie, and civic virtue—because we rarely engage together in civic practices. Within the historic tradition of civic republicanism, diverse individuals—not diverse by today's standards, but each self-interested and unique in his own way—became citizens as they engaged together in civic practices. And while multicultural America presents more of a challenge, I believe it would be productive to consider the ways in which engagement together in civic practices today might constitute our diverse peoples as American citizens in a substantive, participatory republican sense.

Traditionally, military service played a key role in the constitution of republican citizenship, because it was military service that, despite its sometimes deadly purposes, instilled in individuals the virtues necessary for self-government aimed at the common good—selflessness, courage, camaraderie, patriotism, and civic virtue. In arguing for the reconstitution of the Citizen-Soldier tradition, one of the changes I call for involves a shift in emphasis from *military* service to military *service*. On this point I join other democratic theorists who advocate a modern-day transformation of the Citizen-Soldier tradition into a program of national or civic service. For example, Benjamin R. Barber argues that "universal citizen service could offer many of the undisputed virtues of military service: fellowship and camaraderie, common activity, teamwork, service for and with others, and a sense of community." But while citizen service would preserve the virtues of the Citizen-Soldier tradition, it would also minimize its corresponding vices: "In place of military hierarchy, it could offer equality; in place of obedience, cooperation; and in place of us/them conflict of the kind generated by parochial participation, a sense of mutuality and national interdependence."[14] Replacing military service with a broader vision of civic service would facilitate the inclusion of all Americans in the practices constitutive of republican citizenship and would thereby minimize the risks of fusion, homogeneity, and the construction of a totalizing identity.

In addition, my rereading of civic republican theory through the lens of contemporary feminist theory strives to move away from the idea of citizenship as an *identity* and toward a reconceptualization of citizenship as a set of *civic practices*. Because identity always forms in opposition to what it excludes, emphasizing a common *identity* risks exacerbating the vicious side of the Citizen-Soldier tradition—its chauvinism, exclusion, and conformity. In addition, if this deep sense of civic *identity* is produced predominately through *military* service, this makes nationalistic military conquest more likely. In order to wage war, one must strongly identify as a member of a "people" or a "nation," and this type of deep identification most easily develops in opposition to an "enemy" on whom one wages war.

Shifting from an emphasis on *military* to an emphasis on *service* should help minimize the vices inherent in the Citizen-Soldier tradition and help change our definition of citizenship from a common identity to *participation in a set of civic practices*. Civic *service* does not require the same depth of identification, as does *military* service. Participation in a wide variety of civic practices as one part of one's life produces a multidimensional, less totalizing form of identity. Engagement in such civic practices could constitute individuals as American citizens, but not as purely American and nothing else.[15] Moreover, situating military service within a broad array of civic practices should remind us that a democratic society has a military not just to defend its borders but also to defend its democratic principles, including equality and participatory citizenship.

Nevertheless, any *citizenship of civic practices* still runs the risk of creating vices—chauvinism, exclusivity, and enforced conformity—along with virtues—patriotism, camaraderie, and civic virtue. However, it seems to me that the large degree of diversity existing in contemporary America makes the risks connected with the creation of civic bonds worth taking. As Barber puts it,

> the fragmentation and pluralism of most contemporary liberal democratic societies would seem to leave ample room for a safe infusion of communitarian values. . . . Neighborhood ties and the affective bonds that emerge out of common activity are obviously less risky than patriotism, which in modern times has often meant chauvinism or jingoism, and less dangerous than civil religion, which has often spawned a style of fundamentalism zealotry incompatible with the separation of church and state and with genuine pluralism.[16]

The *citizenship of civic practices* inherent in the Citizen-Soldier tradition requires that individuals participate together in civic practices, if they want to become citizens. Only acting together as citizens can instill in us the affective bonds that form the necessary prerequisite for attending to what we have in common rather than what divides us.

Citizen-Soldiers and Manly Warriors begins with a pair of chapters that lay the theoretical groundwork for the study by examining the role of the Citizen-Soldier in the work of Niccolo Machiavelli and Jean-Jacques Rousseau.

Together these chapters provide a coherent picture of a complicated tradition that contains at its center a performative understanding of citizenship, which I term the *citizenship of civic practices.* At the same time, however, this juxtaposition also reveals two distinctly different versions of a common tradition that produce very different sets of political problems.

Chapter 2, "Machiavelli and the *Citizenship of Civic Practices,*" begins with an examination of the Citizen-Soldier ideal in the work of Machiavelli, the first modern theorist of civic republicanism. The chapter argues that the Citizen-Soldier plays an absolutely essential role in Machiavelli's oeuvre because this figure embodies the linkages between a series of ostensibly contradictory ideals: the civic realm and the militia, participatory citizenship and *armed masculinity,* civic virtue and *virtu,* republican ideals and militarism. After a thorough discussion of the virtues and vices inherent in the tradition as a whole, the chapter concludes that because Machiavelli emphasizes the importance of external enemies in creating unity among citizens, he ends up moving away from the virtue of the civic militia as a defensive force and toward the vices of imperialism and conquest.

Chapter 3, " 'Jean-Jacques . . . You Are a Genevan': Civic Festivals, Martial Practices, and the Production of Civic Identity," examines the major eighteenth-century proponent of the Citizen-Soldier. Although Rousseau's political theory of the Citizen-Soldier parallels Machiavelli's to some extent, it also entails several interesting differences. The chapter argues that the Citizen-Soldier stands at the very center of Rousseau's entire theoretical framework because it embodies a set of practices that produce the necessary foundation for republican self-rule: The civic and martial practices constitutive of the Citizen-Soldier also produce patriotism, fraternity, and civic virtue, as well as the *general will* itself, the essential prerequisites for government aimed at the common good. At the same time, however, the practices of the Citizen-Soldier tradition also produce a set of vices that form the flip sides of those same virtues: patriotism can become nationalism, fraternity can lead to a totalizing civic identity, civic virtue can yield fusion, and the *general will* can produce homogeneity.

Chapter 3 concludes that while these interrelated virtues and vices are often mutually constituted, Rousseau's version of the Citizen-Soldier actually exacerbates the vicious side of this tradition because he creates unity through an all-encompassing set of civic and martial practices, the channeling of all passion toward the fatherland, and the production of a totalizing civic identity that replaces all others. That is to say, while he avoids the problems associated with the creation of unity through opposition to an external enemy—Machiavelli's vices of imperialism and conquest—Rousseau creates unity by forging versions of fraternity, patriotism, and civic identity that are so strong that they slip easily into fusion, nationalism, and homogeneity.

After constructing a political theory of the Citizen-Soldier, I turn to a historically grounded examination and analysis of the civic and martial practices generated by this tradition in eighteenth- and nineteenth-century America. Chapter 4, "The Civic Rituals of the American Citizen-Soldier," examines a third, less theoretical version of the Citizen-Soldier tradition that developed within the context of American political history. A central figure within the American tradition of civic republicanism, the Citizen-Soldier constituted a cultural and political ideal that was more important for the production of masculine citizens than for actual military effectiveness. And as we would expect based on our previous theoretical analysis, the American rendition of the Citizen-Soldier also produces an interrelated set of virtues and vices: While civic and martial practices create citizen-soldiers and instill in them patriotism, civic virtue, and fraternal solidarity, these same practices can also yield xenophobia, racism, violence, and homogeneity. Thus, the *citizenship of civic practices* has antidemocratic as well as democratic potential, and these two oppositional impulses are directly related.

Chapter 4 traces the history of the civic militia in American political history and concludes with an analysis of the *national* draft that was established in the early twentieth century. I argue that despite its emphasis on the military obligations of all male citizens, the Selective Draft Act of 1917 does *not* in fact represent the epitome of the Citizen-Soldier tradition, because it does not link military service to participatory citizenship. As I argue throughout this book, the Citizen-Soldier constitutes a *normative* ideal that links military service to participatory citizenship; it is not merely an empirical description of a military made up of liberal citizens. Nor can it be reduced to the idea that citizens serve in the military only temporarily. Rather, the Citizen-Soldier tradition makes soldiering central to the process of *becoming* a citizen, because martial practices instill in citizens the virtues *required* for participation in self-government aimed at the common good. Thus, because the Selective Draft Act did not connect the national draft with a reworked version of participatory citizenship suited to America's new status as a nation-state, it does not fully exemplify the political theory of the Citizen-Soldier.

Finally, the book brings its theoretical and historical discussion to bear on two important contemporary political phenomena in American politics. In chapter 5, "Citizen-Soldiers, Blood Brothers, and the New Militias: Interrogating the Republican Discourse of the American Right," I examine the use of the rhetoric of the Citizen-Soldier tradition by the "New Militia movement." The chapter argues that America's racist and anti-Semitic right-wing groups have been able to build support for their agenda by "masking" their antidemocratic claims with two democratic discourses characteristic of the American political tradition— "identity politics" and the Citizen-Soldier tradition. In so doing, they have rendered their ideas more palatable to those who do not want to be known as quasi-fascists, in particular, the "angry white men" who feel scared, economically and politically threatened, and shut out of the American political process. I ultimately conclude that the New Militia movement embraces the American Citizen-Soldier tradition only in its most vicious form; it feeds on and reproduces the vices of xenophobia, racism, violence, and homogeneity, while it rejects the normative democratic principles that form an essential component of the political theory of the Citizen-Soldier. Therefore, if we consider the *political theory* of the Citizen-Soldier in its best *democratic* sense, then we must reject the claim of antidemocratic groups that they stand within that civic republican tradition.

Finally, chapter 6, "Troubling *Armed Masculinity:* Military Academies, Hazing Rituals, and the Reconstitution of the Citizen-Soldier," begins by examining the recent struggle of women for admission into the Citadel and the Virginia Military Institute (VMI), two institutions that self-consciously claim to be creating "citizen-soldiers." After reviewing the feminist literature on gender and the military and laying out my version of performativity theory, the chapter examines why these two military academies believe that the presence of women would completely disrupt their educational methods. What my study reveals is that within the Citadel and VMI, as well as within the armed forces more generally, misogynistic and homophobic hazing rituals still play a central role in the process through which a bunch of male individuals are transformed into a fraternity of manly warriors. After analyzing those findings in light of the political theory of the Citizen-Soldier, I conclude that this unnecessary and antidemocratic method of creating soldiers actually threatens to undermine the very democratic ideals citizen-soldiers are primarily charged with defending.

Chapter 6 concludes with a plan for reconstituting the Citizen-Soldier tradition in light of contemporary commitments to gender equality. In order to make republican citizenship truly democratic, we must sever the link between manhood and citizenship that has historically characterized this tradition. Using my concept of performativity, I argue that the counterhegemonic engagement of "women" in the practices traditionally constitutive of masculine citizen-soldiering combined with a commitment to democratic ideals should yield a new form of civic identity that is not constructed through the exclusion of biological females. Gender parity in the American military is important, but not enough. If we want to reconstitute the Citizen-Soldier tradition in America, we need to change the type of masculinity produced by the military, reintroduce the military to its civic purposes, expand the *citizenship of civic practices* to include other, nonmartial forms of service, and give citizens a greater role in political decision making. Only then will we have actualized the democratic potential inherent in the Citizen-Soldier tradition.

Notes

1. Wendy Brown, *Manhood and Politics: A Feminist Reading in Political Theory* (Totowa, NJ: Rowman & Littlefield, 1988).

2. For Butler's discussion of performativity, see *Gender Trouble: Feminism and the Subversion of Identity* (New York: Routledge, 1990); *Bodies that Matter: On the*

Discursive Limits of "Sex" (New York: Routledge, 1993); and her essays in *Feminist Contentions: A Philosophical Exchange,* ed. Seyla Benhabib, Judith Butler, Drucilla Cornell, and Nancy Fraser (New York: Routledge, 1995).

3. For a discussion of this distinction in the German context, see Manfred Steger and F. Peter Wagner, "Political Asylum, Immigration, and Citizenship in the Federal Republic of Germany," *New Political Science* 24/25 (Summer 1993), 65.

4. Butler, *Gender Trouble,* 6.

5. Butler, *Gender Trouble,* 34.

6. Butler, *Bodies that Matter,* x.

7. As I define it, postmodernism includes the wide diversity of theories that contest the basic premises of political modernity. This would include Butler's work, although she personally rejects the term *postmodernism,* which she considers too monolithic, in favor of the term poststructuralism. See Butler, "Contingent Foundations," in *Feminist Contentions,* 35–39.

8. Judith Butler, *Bodies that Matter,* 95.

9. Butler, *Bodies that Matter,* 95; *Gender Trouble,* 142.

10. Butler, *Bodies that Matter,* 125.

11. Seyla Benhabib, "Feminism and Postmodernism," in *Feminist Contentions,* 20.

12. Benhabib, "Feminism and Postmodernism," 21.

13. Martha Nussbaum, "The Professor of Parody: The Hip Defeatism of Judith Butler," *The New Republic,* 22 February 1999, 43.

14. Benjamin R. Barber, *Strong Democracy: Participatory Politics for a New Age* (Berkeley: University of California Press, 1984), 302. For a realistic and detailed plan for the "reconstruction of the citizen-soldier," see Charles C. Moskos, *A Call to Civic Service: National Service for Country and Community* (New York: The Free Press, 1988).

15. For an interesting discussion of this idea, see Michael Walzer, *What It Means to Be an American: Essays on the American Experience* (New York: Marsilio Publishers, 1996).

16. Barber, *Strong Democracy,* 243.

Chapter 2

Machiavelli and the
Citizenship of Civic Practices

There cannot be good laws where armies are not good, and where there are good
armies, there must be good laws.

—Machiavelli, the *Prince*

Niccolo Machiavelli's work forms a quintessential example of the Citizen-
Soldier ideal in civic republicanism. Within the civic republican tradition, the
Citizen-Soldier ideal is absolutely central for several reasons. In the first place,
it links the two realms in which a republic must remain free and autonomous:
It links the civic realm, in which republican citizens govern themselves for the
common good, with the civic militia, through which citizen-soldiers protect
their liberty and autonomy from the threat of external enemies. As we shall
see, the Citizen-Soldier constitutes a normative ideal that necessarily entails a
commitment to a set of republican political principles, including liberty, equality, fraternity, the rule of law, the common good, civic virtue, and participatory citizenship.

The Citizen-Soldier ideal forms the centerpiece of what I call a *citizenship
of civic practices*. According to this model, individuals actually *become* citizens as they participate together in civic practices, traditionally including
those of the civic militia. More specifically, within the civic republican tradition, "citizen" is not a prepolitical identity. Individuals are not "citizens" simply by virtue of the fact that they live within certain borders (*ius solis*) or because they have a particular class or ethnic heritage (*ius sanguinis,* or what
could be called the *citizenship of blood*). Instead, engagement in civic practices *produces* a common civic identity; it constitutes diverse individuals as
citizens. Never finally achieved, citizenship must be constantly constructed
and reconstructed through engagement in civic practices. And traditionally,
participation in the civic militia formed the main practice through which citizenship was constructed.

15

Both virtues and vices characterize the *citizenship of civic practices*. On the one hand, through participation in the civic militia, individuals become citizen-soldiers as they learn patriotism, selflessness, and fraternity, all of which coalesce into civic virtue. On the other hand, these same martial practices can teach citizen-soldiers the vices that form the flip sides of these same virtues: Patriotism becomes conquest, selflessness conformity, and fraternity chauvinism. And instead of civic virtue, we get the other half of Machiavellian *virtu:* a combative *armed masculinity*. As we shall see, within the Citizen-Soldier tradition, the creation of participatory citizenship historically entails both an enemy and the denigration of femininity. While any version of the *citizenship of civic practices* will produce vices as well as virtues, when service in the civic militia forms the primary civic practice constitutive of citizenship, the vices are more prominent. Because of the centrality of the Citizen-Soldier ideal to Machiavelli's vision, his republican virtues are inextricably linked to a corresponding set of vices.

The Republican Reading of Machiavelli

The argument that the Citizen-Soldier ideal stands at the very center of Machiavelli's theoretical framework depends upon a republican reading of Machiavelli. However, not all political theorists see Machiavelli as a republican theorist.[1] The debate over the political orientation of Machiavelli's work grows out of the apparent contradiction between the autocracy of the *Prince* and the republicanism of the *Discourses*. Those who focus primarily on the *Prince* doubt Machiavelli's commitment to republicanism. For example, Leo Strauss and Harvey Mansfield consider Machiavelli a "teacher of evil."[2] Other scholars see Machiavelli as an advocate of imperialism and conquest[3] and/or as a protofascist.[4] Many great German thinkers, such as Fichte, Hegel,[5] Herder, Ranke, and Meinecke, stress the role the *Prince* played—for better or for worse—in the emergence of nationalism during the nineteenth century.[6] Often critics of Machiavelli bolster their claims by summoning up the long history of outrage over Machiavelli's work.[7] Still another school of thought stresses that regardless of his intent, Machiavelli divorced politics from morality and so in this way ended up justifying pure power politics, *realpolitik*.[8] Some believe Machiavelli did not actually advocate power politics but simply presented a technical study of how politics works.[9] Others argue that he was in fact tortured over the necessity of doing evil for the sake of good.[10] Nevertheless, in opposition to these views of Machiavelli, an increasingly huge body of scholarship emphasizes the strong republican themes present in Machiavelli's oeuvre, particularly in the *Discourses*.[11] These readers of Machiavelli explain the *Prince* in a variety of ways.[12] The fact that Machiavelli was a lifelong advocate of civic republicanism in practice provides additional evidence for many republican readings of his theoretical work.

My reading of Machiavelli builds on the large body of scholarship that portrays him as a republican theorist, and this chapter discusses Machiavelli's republican ideals in depth. My focus on the Citizen-Soldier ideal in Machiavelli's work bolsters arguments that see the *Prince* as providing instructions that, if followed, would lay the groundwork for the transition from a monarchy to a republic.[13] This reading relies upon the famous last chapter of the *Prince,* in which Machiavelli states the following:

> If then, your glorious family resolves to follow the excellent men I have named who redeemed their countries, she must *before all other things, as the true foundation of every undertaking, provide herself with her own armies,* because there cannot be more faithful or truer or better soldiers. And though each one of them is good, they will become better if united, when they see themselves commanded by their own prince and by him honored and maintained. It is necessary, therefore, for her to prepare such armies in order with Italian might to defend herself against foreigners.[14]

Here Machiavelli advises the prince to arm his subjects. In so doing, he advises the prince to lay the foundation for a republic because—as we shall soon see— Machiavelli considered engagement in martial practices as constitutive of republican citizenship and thus as the foundation for a republic.

My ultimate claim is that Machiavelli presents a dialectical vision in which republican ideals and the heroic ethic are reconciled in the figure of the Citizen-Soldier. That is to say, while ostensibly contradictory, Machiavelli's republican citizenship and his emphasis on the heroic ethic—glory, grandeur, and conquest—come together in the figure of the Citizen-Soldier and form a package of interconnected virtues and vices. My reading builds on the work of Mark Hulliung, who begins to get at this dialectic in his book *Citizen Machiavelli:* "If, as we have argued, a Machiavellian potentiality always inhered in the republican tradition, the secondary literature errs in dwelling solely on the 'idealism' of civic humanism and in contrasting it with the so-called 'realism' of Machiavelli."[15] According to Hulliung's rendition, the civic humanists were always champions of the heroic ethic of glory.[16] Hulliung seems to be proffering a dialectical reading when he argues that Machiavelli insisted "that republics and conquest go hand in hand." Criticizing the republican readings of the Florentine, Hulliung argues that

> it is not enough to bridge the gap between the *Prince* and the *Discourses* or to point to Machiavelli's republican progeny in order to make a case for an un-Machiavellian Machiavelli. At this point the standard interpretation of Machiavelli ends when it is precisely at this point that it should begin. Why did Machiavelli favor republics over monarchies? *If the answer may be phrased in terms of liberty, it may equally well be phrased in terms of power,* for his constant principle is that the greatest triumphs of power politics are the monopoly of free, republican communities. The standard scholarly interpretation of Machiavelli is therefore

revisionist; it deletes all that is most striking and shocking in his thought; it is Machiavelli expurgated.[17]

However, Hulliung backs away from a dialectical reading. That is, while Hulliung rightly restores the heroic ethic of heroism, glory, and conquest to the center of Machiavelli's work, in so doing he shortchanges the civic republican aspects, which, as I will demonstrate, are equally central.[18] What Hulliung ultimately does is subordinate Machiavelli's advocacy of republicanism to his desire for *grandezza* and *gloria*. What I am suggesting, on the other hand, is that for Machiavelli the two sets of ideals are equally important and that he synthesizes them through his articulation of the Citizen-Soldier.

Within Machiavelli's oeuvre the heroic ethic and the commitment to civic republican principles—liberty, equality, fraternity, the rule of law, the common good, civic virtue, and participatory citizenship—come together in the figure of the Citizen-Soldier. That is to say, the benefits of republicanism can be obtained only for citizens of a particular republic, and these individuals must constitute themselves as citizens in opposition to an enemy against which they prepare to fight. In other words, to the extent that the Citizen-Soldier ideal forms the foundation of civic republicanism, this tradition presents a framework in which its virtues—including patriotism, selflessness, fraternity, civic virtue, and participatory citizenship—are intertwined with its vices—conquest, conformity, chauvinism, *armed masculinity,* and exclusion. Thus Hannah Pitkin is correct when she states that Machiavelli is "both a republican and something like a protofascist."[19] While the interconnection of virtues and vices always exists within civic republicanism, it exists to a much greater extent when martial practices are privileged over other possible forms of civic practices.

Uniting the Republic in Theory

The ideal of the Citizen-Soldier stands at the very center of Machiavelli's republican theory because it unifies his political understanding, which, in Pitkin's words, "consists of a set of syntheses holding in tension seemingly incompatible truths along several dimensions." The figure of the Citizen-Soldier embodies the linkages between the civic realm and the militia, citizenship and *armed masculinity,* civic virtue and *virtu,* republican ideals and militarism. Pitkin argues that many of the apparent contradictions in Machiavelli's thought come from the unraveling of these syntheses. However, "even when he loses the syntheses," she argues, "he is a better teacher than many a more consistent theorist, because he refuses to abandon for very long any of the aspects of the truth he sees."[20]

The Citizen-Soldier forms the linchpin in Machiavelli's dialectical edifice. Firstly and most democratically, the Citizen-Soldier fuses the militia to the civic realm of republican self-rule. The soldier who risks his life to defend the republic is also the citizen who participates in forming laws for the common good. Both halves of the Citizen-Soldier ideal are equally important: Citizen-soldiers fight to defend their ability to govern themselves for the common good through the rule of law. In other words, the Citizen-Soldier ideal does not mean simply that citizens comprise the military. Normative rather than empirical, the Citizen-Soldier embodies a commitment to civic republicanism, complete with all its ideals: liberty, equality, fraternity, the rule of law, the common good, civic virtue, and participatory citizenship.

Secondly, the Citizen-Soldier ideal represents the fusion of *armed masculinity* onto republican citizenship. That is to say, service in the civic militia plays a key role not only in the constitution of republican citizenship but also in the construction of *armed masculinity,* of what it means to be a man. Through engagement in martial practices, individuals become citizen-soldiers, as they acquire *virtu,* a central concept in Machiavelli's work and one that crystallizes the traditional fusion of masculinity onto citizenship within civic republican tradition. *Virtu* has two meanings in Machiavelli's oeuvre. In the first place, it means civic virtue, the placing of the common good before individual self-interest, a necessary prerequisite to republican self-rule. At the same time, however, *virtu* means the virile action necessary to the domination of *fortuna,* action which, as we will see, constitutes a combative form of *armed masculinity* formed in opposition to a denigrated femininity. Service in the militia teaches *virtu* in both its senses.

Thirdly, the Citizen-Soldier, exemplar of *virtu,* embraces a form of citizenship that is simultaneously republican and militaristic. The republican Citizen is also the Soldier, and every Soldier requires an enemy against which he must prepare to fight. The militarism inherent in the ideal of the Citizen-Soldier plays a key role in unifying the republic as it prepares to defend itself against external enemies and consequently helps prevent the emergence of internal factions. It provides a venue through which citizens of superior ability can serve the republic. And it plays a vital role in the production of *armed masculinity.*

Linking "Good Laws and Good Armies"

The Citizen-Soldier ideal connects the civic realm of legislation to the civic militia, a connection Machiavelli emphasizes in his famous demand for both "good laws and good armies." Both "good laws"—aimed at the common good and created through the participation of citizens—and "good armies"—made up of all citizens and organized as a civic militia—are necessary to the creation and maintenance of a republic. In both the *Prince* and the *Discourses* Machiavelli stresses that "good laws and good armies" are "the principal foundations of all states"—princedoms as well as republics: "And because there cannot be good laws where armies are not good, and where there are good armies, there must be good laws,

I shall omit talking of laws and shall speak of armies."[21] While I would insist that civic participation in legislation is no less important to Machiavelli's republicanism than is participation in the civic militia, taking Machiavelli at his word I begin with his discussion of the necessity of good armies.

Machiavelli emphasizes the need for good armies for several reasons. In the first place, the continued existence of a republic depends quite directly on good armies. That is to say, every state needs good armies because—at least in Machiavelli's world—all states are vulnerable to attack from competing states.[22] Machiavelli knew very well that a republic is a very fragile entity that must be carefully nurtured and defended. Although for the first third of the fifteenth century "Florence was a genuine republic"—albeit one that restricted citizenship to an elite group of wealthy, powerful men[23]—by the time Machiavelli wrote his political theory of republicanism, the Florentine republic was merely a memory.[24] Thus, he recognized the precariousness of republican government and its vulnerability to both external and internal threats. Only an armed state can protect itself from foreign conquest and thus maintain its republican ideals.

In fact, many scholars emphasize that military threats from northern Italy and France played a key role in the reemergence of the theory of civic republicanism during the fifteenth century. Although Florence had a tradition of self-government, it was not until the fifteenth century that theoretical justification of republicanism began to emerge. By that point the Florentine republic was being threatened by princedoms and dukedoms to the north. As Hans Baron has demonstrated, around the year 1400 this threat to their way of life led Florentines to begin to think self-consciously about their political practices and to define themselves in connection with the ancient republics rather than with the Empire.[25] In other words, civic republicanism's re-emergence as a theory came out of the Florentine attempt to solidify its identity in the face of the threat of external enemies.

Secondly, for Machiavelli good military organization means a civic militia made up of all citizens: "An army evidently cannot be good if it is not trained, and it cannot be trained if it is not made up of your subjects. Because a country is not always at war and cannot be, she must therefore train her army in times of peace, and she cannot apply this training to other than subjects, on account of the expense."[26] History reveals, he argues, that disarming the people leaves states vulnerable to conquest.[27] And border guards are not enough. States that "make some little resistance on their boundaries" have "no recourse" when

> an enemy has passed them. . . . And they do not see that such a way of proceeding is opposed to every good method. The heart and the vital parts of a body should be kept armored, and not the extremities. For without the latter it lives, but when the former is injured, it dies; and these states keep their hearts unarmored and their hands and feet armored. What this error has done to Florence has been seen and is

seen every day; and when an army passes her boundaries and comes within them close to her heart, she has no further resource.[28]

Thus, the defense of the republic absolutely requires an armed populace.[29]

Moreover, a civic militia made up of all citizens helps maintain peace, preserve liberty, and minimize the possibility of tyranny. Hired mercenaries or foreign auxiliary armies cannot be trusted to protect a state of any type: "mercenary forces never do anything but harm."[30] In the first place, Machiavelli argues that a civic militia has no interest in continuing a war unnecessarily. Professional soldiers do. They "are obliged either to hope that there will be no peace, or to become so rich in time of war that in peace they can support themselves."[31] Because they are not professionals, citizen-soldiers do not expect anything from war, "except labor, peril, and fame." Instead of wanting to remain at war, they wish "to come home and live by their profession." In Machiavelli's words, the citizen-soldier "when he was not soldiering, was willing to be a soldier, and when he was soldiering, wanted to be dismissed" (576). A citizen-soldier "will gladly make war in order to have peace," but "will not seek to disturb the peace in order to have war" (578). Thus the ideal of the Citizen-Soldier should decrease the chances of war not increase them.

The use of professional soldiers also puts the republic at risk of being tyrannized by them. When wars are finished, mercenaries and auxiliaries exist by "exacting money from the cities and plundering the country" (574). On the other hand, Machiavelli argues,

> no great citizen ever presumed . . . to retain power in time of peace, so as to break the laws, plunder the provinces, usurp and tyrannize over his native land and in every way gain wealth for himself. Nor did anybody of low estate dream of violating his oath, forming parties with private citizens, ceasing to fear the Senate, or carrying out any tyrannical injury in order to live at all times by means of warfare as a profession. (575–76)

Moreover, when professional soldiers become tyrants, an unarmed citizenry has no recourse.

Furthermore, not only do mercenaries and auxiliary armies pose the threat of plunder, tyranny, and perpetual war, they also make bad soldiers because the only true inspiration for fighting is the protection of one's own liberty. Mercenaries and auxiliaries "are useless and dangerous; . . . they are disunited, ambitious, without discipline, disloyal." While mercenaries are "valiant among friends, among enemies [they are] cowardly." Consequently,

> in peace you are plundered by them, in war by your enemies. The reason for this is that they have no love for you nor any cause that can keep them in the field other than a little pay, which is not enough to make them risk death for you. They are eager indeed to be your soldiers as long as you are not carrying on war, but when war comes, eager to run away or to leave.[32]

There is a tremendous difference, Machiavelli argues, between "an army that is satisfied and fights for its own glory and an army that is ill disposed and fights for some leader's ambition."[33] Men will willingly and courageously risk their lives only to defend their own liberty:

> Nothing made it harder for the Romans to conquer the people around them and part of the lands at a distance than the love that in those times many peoples had for their freedom, which they defended so stubbornly that never except by the utmost vigor could they be subjugated. We learn from many instances in what perils they put themselves in order to maintain or regain that freedom, and what revenge they wreaked on those who took it from them.[34]

J. G. A. Pocock puts it nicely when he says, "the paradox developed in Machiavelli's argument is that only a part-time soldier can be trusted to possess a full-time commitment to the war and its purposes."[35] Thus only a civic militia can be relied upon to defend a republic.

It is important to note here that when citizen-soldiers fight to defend their republic and their ability to govern themselves through the formation of manmade laws, they fight for a secular political order. Civic republicanism emerged in the fifteenth century in direct opposition to a Christian worldview and political order.[36] As Pocock explains, the revival of Aristotelianism and the revaluation of history led to a break with the medieval Christian "scholastic-customary" framework and a rediscovery of citizenship. The re-emphasis on the importance of time and the deliberative and creative powers of the human mind in both the intellectual movement of civic humanism and the political movement of civic republicanism constituted an attack on the medieval Christian worldview, with its traditional, hierarchic view of society, and on the structures of monarchy and aristocracy it justified: Citizenship requires liberty rather than subjection to tradition, equality rather than hierarchy and rank, fraternity rather than paternity and filiality, and autonomy rather than obedience to natural God-given law and dependence upon natural superiors. Pocock states it baldly: "Machiavelli unequivocally prefers the republic to revealed religion."[37] Citizen-soldiers fight to protect their secular political order and civic ideals. They do not fight for God and His revealed Truth.

Creating *Virtu:* A Common Good for Manly Citizen-Soldiers

Besides being necessary to the continual existence of a republic, the practices of the civic militia are absolutely essential to Machiavelli's *citizenship of civic practices* because they play a key role in the *creation* of masculine citizen-soldiers out of male individuals. That is, for Machiavelli masculinity requires soldiering, and soldiering must be linked to citizenship. Participation in martial practices simultaneously constructs all three characteristics. The interrelated constructions of masculinity, soldiering, and citizenship come together in Machiavelli's concept of *virtu,* which is produced directly by engagement in martial practices.

The Citizen-Soldier ideal embodies Machiavelli's concept of *virtu,* a concept that connects the civic realm to the civic militia and fuses *armed masculinity* to citizenship. *Virtu* has a dual meaning in Machiavelli's work, and citizen-soldiers must possess *virtu* in both senses.[38] In the first place, civic republicanism requires a sense of *civic virtue,* defined as the characteristic whereby individuals place the common good ahead of individual self-interest.[39] In Quentin Skinner's words,

> a self-governing republic can only be kept in being . . . if its citizens cultivate that crucial quality which Cicero had described as *virtus,* which the Italian theorists later rendered as *virtu,* and which the English republicans translated as *civic virtue or public-spiritedness.* The term is thus used to denote the range of capacities that each one of us as a citizen most needs to possess: *the capacities that enable us willingly to serve the common good, thereby to uphold the freedom of our community,* and in consequence to ensure its rise to greatness as well as our own individual liberty.[40]

This is not to say that the common good stands opposed to individual interests. To the contrary, by definition the common good includes the good of each individual.[41] However, government for the common good does constitute an alternative both to a system of rule based on balancing individual interests (such as liberalism) and to rule based on one particular interest that stands opposed to the common good (such as tyranny).

Within the Citizen-Soldier tradition, service in the civic militia plays a key role in the creation of civic virtue, a necessary prerequisite to the willingness to make laws aimed at the common good. According to Skinner, "a leading theme of Book II of Machiavelli's *Discorsi*" is that "the martial virtues," including "courage and determination to defend [the] community against the threat of conquest and enslavement by external enemies," constitute the capacities citizens need to possess in order to uphold the common good, ensure greatness, and protect the liberty of both the community and the individuals who comprise it.[42] Participation in the civic militia requires soldiers to act together for the common good and to sacrifice particular goods to universal ends. In this way military service forms a type of civic education that teaches individuals to act together for the common good during civic legislation. And in this way civic and martial virtue are interconnected. As Pocock explains, "[I]t may be through military discipline that one learns to be a citizen and to display civic virtue."[43] In other words, soldiering privileges certain *virtues* that become attached to citizenship—among them courage, selflessness, fraternity, and patriotism. These virtues force the citizen to rise above his own particular interests and to think of the good of the community as a whole. Thus, within the Citizen-Soldier tradition in general and in Machiavelli's work in particular, martial virtue plays a central role in the construction of civic virtue.

But while civic virtue grows out of martial virtue, martial virtue necessarily presupposes civic virtue: The willingness to self-sacrifice presupposes an identification with the republic. As Pocock argues, the citizen's desire to defend his life in the republic guarantees he will be virtuous in battle.[44] Citizen-soldiers fight to defend their liberty, equality, fraternity, their laws aimed at the common good, and their participatory citizenship. Only the love for one's *patria* and the ideals it represents allows for the possibility of self-sacrifice. Thus, for Machiavelli, the civic militia with its martial virtue is inextricably linked to the realm of civic legislation with its civic virtue by the ideal of the Citizen-Soldier.

The second meaning of *virtu* in Machiavelli's work is *virile political action,* which is directly related to traditional understandings of *masculinity.* As Pitkin explains, *virtu* means "energy, effectiveness, virtuosity" and "derives from the Latin *virtus,* and thus from *vir,* which means 'man.' *Virtu* is thus manliness, those qualities found in a 'real man.' "[45] Machiavellian *virtu* connotes a form of *armed masculinity* that stands opposed to "*effeminato* (effeminate), . . . one of his most frequent and scathing epithets."[46] The republican citizen characterized by *armed masculinity* acquires *virtu* as he battles *fortuna,* a concept Machiavelli understood as feminine.[47] To quote Machiavelli:

> I conclude then (with Fortune varying and men remaining stubborn in their ways) that men are successful while they are in close harmony with Fortune, and when they are out of harmony, they are unsuccessful. As for me, I believe this: it is better to be impetuous than cautious, because *Fortune is a woman and it is necessary, in order to keep her under, to cuff and maul her.* She more often lets herself be overcome by men using such methods than by those who proceed coldly; *therefore always, like a woman, she is a friend of young men, because they are less cautious, more spirited, and with more boldness master her.*[48]

Feminine fortune can be mastered only by an *armed masculine virtu.*

For Machiavelli, engagement in virile martial practices is necessary for the construction of *armed masculinity.* In other words, *armed masculinity* does not exist naturally in male individuals. Instead, it must be *produced.* "Pondering, then, why it can be that in those ancient times people were greater lovers of freedom than in" his times, Machiavelli concludes that the difference comes "from the same cause that *makes men now less hardy.*" That is to say, "this [Christian] way of living, then, *has made the world weak and . . . effeminate.*" Christians are effeminate because they do not engage in virile martial practices. On the other hand, pagans were manly because they were "fiercer in their actions" than the Christian males. Pagan sacrifices were "magnificent, . . . full of blood and ferocity. . . . [And] this terrible sight *made the men resemble it.*" In contrast to this, Christianity has made men effeminate by "glorif[ying] humble and contemplative men rather than active ones."[49] Clearly, Machiavelli did not consider *armed masculinity* a naturally occurring characteristic of male individuals. To the contrary, it must be constructed through participation in the fierce, bloody, and magnificent actions required during military service. Furthermore, Machiavelli constructs his manly *virtu* not only through struggle against a *fortuna* considered feminine but also in opposition to a Christianity also considered feminine. In other words, *virtu* is both manly and secular. As Hulliung puts it,

> Arrayed on one side are the pagan virtues: *virtus,* glory, grandeur, magnificence, ferocity, exuberance, action, health, and manliness; on the other side are the Christian virtues, humility, abjectness, contempt for human things, withdrawal, inaction, suffering, and disease—and the upshot of these Christian " 'virtues,' " he concludes, is the womanish mankind of postclassical times, whose histories are as ignoble as Rome's was noble.[50]

Furthermore, Machiavelli constructs his manly *virtu* not only through struggle against a *fortuna* considered feminine but also in opposition to a Christianity also considered feminine. In other words, *virtu* is both manly and secular. As Hulliung puts it, in Bonnie Honig's words,

> Machiavelli seek[s] in *virtu* a manly alternative to what [he] describes as the feminizing, enfeebling and immobilizing virtue of Christianity.... *Virtu* for Machiavelli is a political excellence, connected with the greatest of all worldly rewards, glory. ... Machiavelli criticizes virtue because its otherworldliness turns men away from the grandest of human worldly endeavors and sabotages the enterprise of politics.[51]

In short, the second meaning of Machiavellian *virtu* is a combative, secular *armed masculinity,* constructed in fierce opposition to femininity.

In laying out the dual meaning of the term *virtu* in Machiavelli's oeuvre, I want to suggest that there is not one "civic *virtu* of Machiavelli's *Discourses*—the excellence of a citizen in a republic"—and a "rather different princely *virtu* of *The Prince,*" as Honig asserts,[52] but rather that the two meanings are unified by the ideal of the Citizen-Soldier. The civic virtue of the citizen and the combatively masculine action of the soldier come together in a figure that exhibits both characteristics at once. Put differently, engagement in the martial practices of the civic militia simultaneously creates citizens with civic virtue, soldiers who display manly *virtu,* and men who acquire their *armed masculinity* in opposition to a denigrated femininity. Consequently, Machiavelli's Citizen-Soldier ideal fuses together soldiering, masculinity, and citizenship.

Because "masculinity" is socially constructed rather than rooted in nature, it can never be secured finally. As a consequence, "femininity"—masculinity's "excess and remainder"[53]—always poses a threat to republican citizen-soldiers. Pitkin lays out four reasons for this. First, the seductive power of young women as sex objects "threatens a man's self-control, his mastery of his own passions." Second, women's erotic power "threatens to infect him with feminine softness." Third, men often succumb to the temptation to violate the chastity of another man's woman, which is one sure way to create political opposition and division. As Machiavelli says, men will tolerate most things as long as they " 'are not deprived of either property or honor,' " and, for him, women constitute both. And fourth, women threaten republican citizenship by "weaken[ing] the manly self-control of citizens . . . [which] tends to privatize the republican citizen, drawing him out of the public square and into the bedroom."[54] In other words, femininity threatens the masculine citizen-soldier's ability to govern himself through legislation aimed at the common good, because it fuels his passions, privatizes him, and disrupts his ability to unite with other men, all of which interfere with the creation of civic virtue. Furthermore, these three things plus the stimulation of feminine softness within him hinder his cultivation of the martial virtues.

Because the Citizen-Soldier constitutes himself in opposition to "femininity," Machiavelli's normative republican vision requires the exclusion of feminine individuals, that is, women.[55] Moreover, it demands that citizen-soldiers stomp out any so-called "feminine" feelings that might exist within themselves. Consequently, as Pitkin argues, Machiavelli

> juxtapose[s] men, autonomy, adulthood, relations of mutuality, politics, the *vivere civile,* human agency in history, and humanness itself, on the one side, to women, childhood, dependence, relations of domination, nature, the power of environment and circumstance, instinct, the body, and animality, on the other. Human autonomy and civility are male constructs painfully won from and continually threatened by corrosive feminine power. Male ambition and human sexuality, however, play ambiguous roles in this struggle, sometimes aiding and sometimes threatening the men. Indeed, the men themselves are ambivalent about the struggle; . . . feminine power seems to be in some sense inside the men themselves. *Only ferocious discipline and terrifying punishments can secure them in the male enterprise of becoming human and autonomous.*[56]

"Masculinity" is never a fait accompli; it must be continually constructed and reconstructed through ferocious military discipline and virile actions. And because of the combative nature of the Citizen-Soldier's masculinity, as Wendy Brown rightly argues, his "construction of manliness . . . entails not mere opposition to but conquest of woman."[57] "Femininity" constitutes a profound danger to a tenuously constructed, *armed masculinity.* Only continual engagement in martial practices can ward off the feminine threat that exists not only outside men but within them as well. In short, the entire structure of Machiavelli's civic republicanism is erected upon the denigration of femininity in all its manifestations.

Soldiering plays a key role in the creation of both *armed masculinity* and citizenship in Machiavelli's work; it is the solder that fuses *armed masculinity* onto citizenship. However, soldiering, Machiavelli tells us, does not come naturally either to men or to citizens. Instead, good soldiers must be *created* through the right institutional context; discipline and training are absolutely essential to this process. Princes and republics that lack their own soldiers "ought to be ashamed," he argues. Using the example of Tullus, Machiavelli argues that a lack of soldiers "comes not from a lack of men fit for warfare but from their own error, because they have failed *to make their men soldierly.*"[58] Tullus' ability was "so great" that "under his direction he immediately made [his men] into very excellent soldiers. So it is truer than any other truth that *if where there are men there are not soldiers, the cause is a deficiency in the prince* and not a deficiency in the position or nature of the country." Warriors, Machiavelli concludes, could exist "in every . . . region where men are born, *if only* there is someone who can *direct them toward soldiership.*"[59] In the *Art of War* Machiavelli says that "ancient examples show that in every country training can *produce* good soldiers, because where nature fails, the lack can be supplied by ingenuity, which in this case is *more important than nature.*"[60] Likewise, in the *Prince* Machiavelli argues that

> if, in so many convulsions in this land and in so much warfare, Italy's military vigor always seems extinct, . . . the cause is that her old institutions were not good, and no one has been wise enough to devise new ones; and . . . in Italy there is no lack of matter on which to impose any form; there is great power in the limbs, if only it were not wanting in the heads.[61]

Clearly, for Machiavelli, the practices of the civic militia play a central role in the actual *production* of manly soldiers.

Soldiering, however, must be connected to citizenship. Machiavelli connects the civic militia directly to the sphere of civic realm in which republican citizens form good laws, when he emphasizes that liberty is the primary underpinning for good armies. "Wherever there are good soldiers," he argues, "there must be good government."[62] This is true because soldiers can be good only when they are protecting their liberty, which can be established and maintained only through good government. That is, Machiavelli believes that liberty must be created through and nurtured in a context of good laws. Good laws both produce and protect liberty by limiting arbitrary power: "To republics, indeed, harm is done by magistrates that set themselves up and by power obtained in unlawful ways, not by power that comes in lawful ways."[63] Good armies require good laws.

Machiavelli argues that only within the context of republican institutions, such as the civic militia and the rule of citizen-authored law, can individuals become citizens. That is, when Machiavelli states that "laws make [men] good,"[64] he argues that outside of the context of law, men can easily slip back into their baser selves:

> As is demonstrated by all those who discuss life in a well-ordered state—and history is full of examples—it is necessary for him who lays out a state and arranges laws for it to presuppose that all men are evil and that they are always going to act according to the wickedness of their spirits *whenever they have free scope.*[65]

That is why Machiavelli favors the rule of law over the rule of men: "Absolute authority in a very short time corrupts the matter and makes itself friends and partisans."[66] For this reason Machiavelli cautions against the long-term delegation of power from the citizens to a magistrate. Although republics should empower a group of citizens to make executive decisions when quick decisions are needed, and so the necessarily slow deliberative process cannot be used, republics must be careful not to delegate this power for long periods of time because "when free authority is given for a long time—that is, for a year or more—it will always be dangerous and will produce good or bad effects according as those to whom it is given are bad or good."[67] And of course whether they are bad or good depends on whether they are able to use power to advance their own particular interests at the expense of the republic.

Partisanship within a republic is problematic because it causes individuals to place private ambition over public good and so compromises the process of legislation for the common good. Citizens in a republic must rule "wholly for the benefit of the state and [should] not in any respect regard private ambition."[68] *Public* citizenship means ruling for the common good, and ruling for the common good constitutes public citizenship. Both are possible only in a republic:

> Without doubt this common good is thought important only in republics, because everything that advances it they act upon, and however much harm results to this or that private citizen, those benefited by the said common good are so many that they are able to press it on against the inclination of those few who are injured by its pursuit. The opposite happens when there is a prince; then what benefits him usually injures the city, and what benefits the city injures him.[69]

Not ruling for the common good causes great disorder, while doing so leads to greatness and increased wealth for the republic.[70]

Machiavelli believes that given the right republican institutional context, human beings can rise above their own narrowly defined self-interests and rule themselves for the common good.[71] Although he holds a cynical view of human nature, as a whole his writings reveal his belief that republican institutions and civic participation can successfully transform selfish individuals into citizens. While he frequently refers to "the nature of men" as "ambitious and suspicious," as unable to "know how to set a limit to its own fortune,"[72] as "insatiable" and therefore always discontented,[73] as shortsighted, vengeful, and ungrateful, he believes that in the context of the rule of law these self-interested men can become republican citizens. Arguing "against the common opinion" that insists on the need for princely rule because "the people, when they are rulers, are variable, changeable, and ungrateful," Machiavelli argues that in the context of the rule of law, the people can rule themselves better than a prince: "A people that commands and is well organized will be just as stable, prudent, and grateful as a prince, or will be more so than a prince, even though he is thought wise." Moreover, "a prince set loose from the laws will be more un-grateful, variable and imprudent than a people. . . . The variation in their actions," he argues, "comes not from a different nature—because that is the same in all men, and if there is any superiority, it is with the people—but from having more or less respect for the laws under which both of them live."[74] Thus, for Machiavelli the key determinant of how men will behave is the context of political institutions within which they live; men's actions are not necessarily driven by "human nature." While without the rule of law, men will interact only on the basis of power and self-interest, under the rule of law, men are capable of governing themselves in accordance with the common good.

In fact, Sebastian de Grazia argues that Machiavelli's contention that naturally selfish individuals can still govern themselves for the common good marks his break with ancient philosophy:

> In our philosopher's world men do not have an inherent impulse toward the common good. Quite the reverse. These wicked and unruly men are not just a few: they comprise mankind. . . . This is Niccolo's third major contribution to political philosophy: the vision of a world in which rational brutes must reach the common good. Binding a permanent, state-prone, or political and social, human nature to the end of a good-in-association, or the common good, was a triumph of ancient political philosophy. Niccolo snaps the link of nature and end. The common good is still the goal but no longer do men reach it naturally.[75]

Born rational brutes, (male) individuals are *capable* of governing themselves for the common good, *but only in the context of participatory republican institutions.*

On this point my reading differs from the one presented by Wendy Brown in *Manhood and Politics.* In contradistinction to Brown, who stresses that Machiavelli's work emphasizes the "immutable characteristics of man,"[76] I would argue that for Machiavelli, man's "second nature" is much more important than any essential human nature. While Brown declares that "Machiavelli harbors no illusions about the usefulness of a political theory based upon 'men as they might be' rather than men as they are or can be," I would argue that Machiavelli does in fact offer a vision of how men could be *given the right republican institutions.* More specifically, Brown uses the following passage to emphasize the "animality" Machiavelli attributes to human nature:

> What great difficulty *a people accustomed to living under a prince* has later in preserving its liberty, if by any accident it gains it. . . . And such difficulty is reasonable because *that* people is none other than a brute beast which, though of a fierce and savage nature, has always been cared for in prison and slavery. Then if by chance it is left free in a field, since it is *not used to* feeding itself and does not know the places where it can take refuge, it becomes the prey of the first one who tries to chain it.[77]

In opposition to Brown's usage, I would stress that this passage emphasizes not man's "immutable" human nature but rather the importance of political institutions in reconstructing the "nature" of man. That is, "people accustomed to living under a prince" have not been transformed into citizens capable of ruling themselves through participation in civic and martial practices. For Machiavelli these practices are critically important, precisely because only they can construct citizens out of ambitious, self-interested individuals.

In other words, Machiavelli espouses a *citizenship of civic practices* in which a man never finally *becomes* a citizen, in the sense that he will always think and act in terms of the common good. Outside of republican institutions—such as the rule of citizen-authored law and the civic militia—which require him to behave as a citizen, he will cease to be one. Hence, the process of becoming a citizen is never finished. Citizens must be constantly re-produced through engagement in civic practices. For Machiavelli, participation in the twin practices of civic republicanism—in both the civic militia and civic legislation—actually *produces* masculine citizen-soldiers out of male individuals. That is why only a republic contains citizens.

This *citizenship of civic practices* contrasts with two other conceptions of citizenship: *ius solis* and *ius sanguinis*—or what I call the *citizenship of blood*. This latter concept bases citizenship on common bloodlines and so to members of a particular ascribed group. With his positive appeal to the integration of new people into a republic—whether by choice or by force—Machiavelli clearly rejects the idea that citizenship should be based on common blood. He is not interested in securing citizenship only for those with Italian blood or noble blood. On the other hand, Machiavelli does not define citizens as any group of individuals living within a particular bounded territory. For instance, individuals living within particular borders but under the rule of a prince are called subjects, not citizens. Hence, Machiavelli does embrace a *citizenship of civic practices* that requires participation in self-rule and in the civic militia. To be a citizen in a civic republic, one must constantly act as a citizen; the category of "citizen" is never finally consolidated.

In other words, implicit in Machiavelli's vision is a performative understanding of both civic and gender identity. Gender and citizenship are not prepolitical categories. That is to say, there are not "men," "women," and "citizens" who then choose whether or not to engage in political action. When Brown argues that "manhood constructs politics," she is arguing that prepolitical, cultural understandings of "manhood" directly affect the shaping of politics because men make politics.[78] In this configuration, manhood pre-exists politics. My analysis is slightly different. Instead of viewing "men," "women," and "citizens" as prepolitical categories, I contend that both citizenship and gender are constituted through engagement in particular practices—civic and martial practices for men and domestic practices for women. Consequently, we cannot simply state with Jean Bethke Elshtain that "Machiavelli's politics eliminates women by definition from the most important field of citizen involvement, military exploits,"[79] for this would assume that for Machiavelli gender identity pre-exists politics. Instead, we must recognize that manhood for Machiavelli is actually constituted through engagement in politics. In other words—turning Brown on her head—politics constructs manhood.

The idea that gender is performatively constructed rather than rooted in nature is important for democratic and feminist theorists because it allows for the possibility of change. Ironically, despite the thoroughly masculine character of Machiavelli's Citizen-Soldier, his implicitly performative understanding of citizenship—his *citizenship of civic practices*—actually allows us to imagine the possibility of including female individuals as citizens. That is, if men were *naturally* more capable of autonomy and mutuality than women, then the possibility that women could ever become autonomous republican citizens would be profoundly problematical. However, this is not Machiavelli's argument. As I have demonstrated, Machiavelli does not argue that men are *naturally* autonomous and capable of mutuality. In fact, Machiavelli repeatedly stresses that it is only within a carefully constructed context of always-fragile republican institutions that men are able to transcend their ambitious, power-seeking, self-interested behavior and learn to become autonomous republican citizens capable of mutuality. And although many scholars have shown that Machiavelli considered only men capable of achieving autonomy and political mutuality, he did not argue that men are *naturally* that way. On the contrary, Machiavelli argues that men become citizens capable of autonomy and mutuality only through participation in civic and martial practices. Furthermore, this constitution of masculine citizens is never finally completed because once the civic republican context is ruptured, men revert back to being self-interested power-seekers. Outside of the practices that produce republican masculinity, "men" become effeminate.[80]

What I am suggesting, then, is that if men's natures are subject to social construction through political practice, then so are women's. Women are not essentially more dependent, natural, and corporeal than men. To the contrary, they remain that way—partly at least—because of exclusion from civic and martial practices. Consequently, the *citizenship of civic practices* contains the democratic potential of including female individuals in republican citizenship: Perhaps female individuals could become republican citizens alongside "men" *if* they began to engage in the same civic and martial practices. At the same time, however, as we will see, the democratic potential of the *citizenship of civic practices* is undermined when the primary civic practice constitutive of citizenship is service in the civic militia, because the martial practices inherent in the civic militia produce a particularly combative form of *armed masculinity* that ultimately undermines the mutuality entailed in the idea of republican citizenship.

Identity Out of Diversity

One of the most democratic aspects of the *citizenship of civic practices* is the construction of politically equal citizens out of diverse individuals. The practices of citizenship assume a certain amount of political equality among those to which it is extended. That is to say, ideally, all citizens should be included in the process of self-rule. In Rome, Machiavelli tells us,

> a Tribune, and any other citizen whatever, had the right to propose a law to the people; on this every citizen was permitted to speak, either for or against, before it was decided. This custom was good when the citizens were good, because it has always been desirable that each one who thinks of something of benefit to the public should be permitted to state his opinion on it, in order that the people, having heard each, may choose the better.

Political equality is essential to civic legislation because power imbalances compromise the possibility of ruling for the common good. In Rome, Machiavelli continues, "when the citizens became wicked" and thus concerned only with their own self-interest, civic legislation "became very bad, because only the powerful proposed laws, not for the common liberty but for their own power, and for fear of such men no one dared to speak against those laws. Thus the people were either deceived or forced into decreeing their own ruin."[81] Thus, without political equality civic legislation cannot occur, and without civic legislation there can be no citizenship.

Machiavelli argues that republican self-rule is superior to autocracy. In comparing the rule of the people to the rule of princes, Machiavelli argues that

> as to judging things, very seldom does it happen, when a people hears two men orating who pull in opposite directions, that if the two are of equal ability, the people does not accept the better opinion and does not understand the truth it hears. And if in matters relating to courage or that seem profitable, as we said above, it errs, many times a prince too errs as a result of his own passions, which are many more than those of the people. It also appears that in choosing magistrates a people makes far better choices than a prince, nor will a people ever be persuaded that it is wise to put into high places a man of bad repute and of corrupt habits—something a prince can be persuaded to do easily and in a thousand ways.[82]

Citizens are more likely to rule for the common good and appoint qualified magistrates and less likely to govern according to passion and whim than are princes.

While Machiavelli stresses the need for *political* equality, however, he does not call for the elimination of all differences. In fact, not only does diversity exist, but it constitutes one of the benefits of republican government: "Thence it comes that a republic, being able to adapt herself, by means of the diversity among her body of citizens, to a diversity of temporal conditions better than a prince can, is of greater duration than a princedom and has good fortune longer."[83] There exist, Machiavelli argues, in every republic "two opposed factions, that of the people and that of the rich."[84] He goes on to insist, moreover, that in Rome it was precisely the differences between the nobility and the people that formed "a first cause" in keeping "Rome free." That is to say, although noisy, he argues, "those dissensions" brought "good effects."[85] More specifically, those dissensions in Rome did not cause "bloodshed" and were not "injurious" because of "honorable conduct" rooted in "good education," which was rooted

> in good laws; good laws in those dissensions that many thoughtlessly condemn. For anyone who will properly examine [the outcome of these dissensions] will not find that they produced any exile or violence damaging to the common good, but rather laws and institutions conducive to public liberty.[86]

In other words, the diversity of views considered when all citizens participate leads to the creation of good laws. And the process of participating in the formation of these laws contributes to the constitution of republican citizens out of diverse individuals. Thus, Machiavelli envisions *political* equality that allows for diversity.[87]

Moreover, governing for the common good does not annihilate individuality, but to the contrary actually creates citizens out of diverse individuals. Grazia argues that Machiavelli's conception of the common good locates "the benefit not on the community considered as an abstract whole, but on its members as individuals (each one) or as superior numerically (the most)."[88] Pitkin too stresses that Machiavelli's common good requires neither "a selfless merging" nor "submission to . . . repressive discipline." Instead, his republic

> offers each Citizen, each class of Citizens, the genuine possibility of fulfilling individual needs, pursuing separate interests, expressing real passions; it does not depend on sacrifice, either voluntary or enforced. Yet the selfish and partial needs, interests, and passions brought into contact with the conflicting needs, interests, and passions of other Citizens and ultimately redefined collectively in relation to the common good—a common good that emerges only out of the political interaction of the Citizens.[89]

The Citizen, she argues, can develop *virtu* only in the "actual experience of citizen participation. Only in crisis and political struggle are people forced to enlarge their understandings of themselves and their interests."[90] Participation in the process of legislating for the common good leads not only to good laws but also constitutes diverse individuals as citizens.

The important point here is that while in Machiavelli's world, republican citizenship was in fact restricted to an elite group of men, within this group differences existed. In other words, although from the outside the group seems very homogeneous—especially from a late-twentieth-century perspective—class differences existed and, moreover, the people included in the group no doubt believed themselves to be a diverse group of unique individuals with often conflicting desires. This view of civic republicanism offers us the possibility of imagining the creation of political equality out of a much greater diversity of individuals and the forging of a *citizenship of civic practices* that does not annihilate differences.

Nevertheless, though Machiavelli insists that differences among citizens contribute to a well-governed republic, he also cautions that these differences can be destructive when they lead to the formation of factions. Because of natural inequalities of ability some men will achieve greater reputations than others. Unless harnessed to serve the public good, men of superior ability could destroy a republic. Reputation should be regulated, so that "citizens will get repute from popularity that aids and does not injure the city and her liberty."[91] In other words, reputation should be gained and honors given for deeds that benefit the common good. Reputations "gained in private ways"—by "conferring . . . benefits on various private persons, by lending them money, marrying off their daughters, protecting them from the magistrates, and doing them similar private favors"—"are very dangerous and altogether injurious" because these acts "make partisans of their benefactors and give the man they follow courage to think he can corrupt the public and violate the laws."[92] Large differences of wealth and power can destroy a republic. In sum, although diversity and natural inequalities in ability can contribute to the health of a republic, these differences can become divisive and lead a republic to ruin. To stave off this possibility, there must be political equality among citizens and publicly acclaimed ways for men with superior abilities to serve the common good. There must also be political equality established and maintained through the rule of law to prevent individuals from exercising arbitrary power.

That is to say, the struggle of diverse individuals to act together for the common good both invigorates and threatens the existence of a civic republic. Honig emphasizes this in her discussion of Machiavelli. As she explains it, Machiavelli stressed that insatiable human desires "cannot be extirpated but they can be held in a creative and productive tension." Only the "perpetuity of [the] struggle" between the nobles and the people "and the institutional obstacles to its resolution, prevent any one party from dominating and closing the public space of law, liberty, and *virtu*."[93] Republicanism necessarily entails struggle and dissension as diverse individuals act together. While this always involves risk, it also keeps a republic vibrant.

Republican Ideals and Militaristic Conquest

Machiavelli argues that preparing for war helps unify citizens for the common good: "The disunion of republics usually results from idleness and peace; the cause of union is fear and war."[94] War facilitates civic republicanism in three ways. In the first place, as I have been arguing, participation in the civic militia creates citizen-soldiers out of diverse individuals. In other words, by serving together in the civic militia, individual males achieve a sense of patriotism, selflessness, and fraternity, and they gain a common civic identity. Preparing for war provides a venue through which individuals can act together for the common good and so become citizens. And the existence of a common enemy helps prevent the formation of factions within a republic. Put differently, preparing for war keeps the focus on what citizens have in common and places the enemy outside of the republic's borders rather than within them.

Secondly, being at war allows men of superior ability both to serve the republic and to achieve personal glory as military officers. Republics, Machiavelli tells us, tend to "show this defect":

> They pay slight attention to capable men in quiet times. This condition makes men feel injured in two ways: first, they fail to attain their proper rank; second, they are obliged to have as associates and superiors men who are unworthy and of less ability than themselves. This abuse in republic has produced much turmoil, because those citizens who see themselves undeservedly rejected, and know that they can be neglected only in times that are easy and not perilous, make an effort to disturb them by stirring up new wars to the damage of the republic. When I consider possible remedies, I find two: the first is to keep the citizens poor, so that, when without goodness and wisdom, they cannot corrupt themselves or others with riches; the second is to arrange that such republics will continually make war, and therefore always will need citizens of high repute, like the Romans in their early days.[95]

As Honig puts it,

> if a republic's energies are not expended in war, they turn inward. If legitimate, institutional avenues of expression are not available, instincts and ambitions will seek other avenues of expression, and the result will be destabilizing conspiracies and the eventual overthrow of the regime.[96]

A civic militia constantly preparing for war helps republics maintain unity by providing a way of rewarding talented men in accordance with the common good. Thus, militarism channels the constant struggle that both invigorates and threatens the republic into service for the common good.

And thirdly, the waging of war is necessary to the construction of Machiavelli's *armed masculinity*. In his words, "[I]f heaven is so kind to [a city] that she does not have to make war, the effect might be that ease would make her effeminate or divided; these two things together, or either alone, would cause her ruin."[97] Soldiering is essential to the constitution of *armed masculinity* for Machiavelli. Soldiering simultaneously produces *armed masculinity* and republican citizenship and melds the two into one.

At first glance, Machiavelli's suggestion that "republics [should] continually make war" seems to contradict one of his main justifications for the Citizen-Soldier ideal. As discussed earlier, Machiavelli argues that one of the important characteristics of the civic militia is that citizen-soldiers are less likely to wage war than professional soldiers and that men of superior ability can cause problems for a republic by "stirring up new wars." Now we see that he also advocates continual preparation for war as a way of preventing the latter problem and unifying the republic: "Because a country is not always at war and cannot be, *she must therefore train her army in times of peace.*"[98] Although citizen-soldiers "will not seek to disturb the peace in order to have war,"[99] they must constantly prepare for war.

To resolve this paradox, Machiavelli uses the ideal of the Citizen-Soldier dialectically to reconcile republican citizenship with militarism. Preparing to fight external enemies imbues citizen-soldiers with patriotism, selflessness, fraternity, civic virtue, and civic participation, as well as *armed masculinity.* Unfortunately, along with these virtues come the vices of conquest, conformity, chauvinism, combativeness, and exclusion. Machiavelli's republican citizens need to prepare for war. With his Citizen-Soldier, Machiavelli attempts to balance republican ideals with the heroic ethic. He does not subordinate one to the other. So while Hulliung mistakenly argues that Machiavelli puts republicanism in the service of conquest, Grazia also errs when he argues that our theorist places conquest in the service of republican ideals.[100] Grazia might be right that "Niccolo is not a militarist at heart," but his Citizen-Soldier is both a republican citizen and a militaristic soldier. Embodying *virtu* in both senses—civic virtue and *armed masculinity*—the Citizen-Soldier synthesizes a variety of oppositional ideals in Machiavelli's work, including civic republicanism and militarism.

Machiavelli values the heroic ideal. That is why, as Hulliung points out,

> of all republics past and present to choose from, it was the world-conquering Roman republic that arrested Machiavelli's attention. The ancient model he admired and hoped to reproduce in modern times was none other than that singularly expansionary, singularly successful Roman republic whose way of life had been the fulfillment of *virtus,* and ethic of glory, grandeur, and heroism.[101]

But at the same time, as Pocock explains, the continued existence of the republic also requires an internal commitment to republican ideals: "The republic can dominate *fortuna* only by integrating its citizens in a self-sufficient *universitas,* but this in turn depends on the freely participating and morally assenting citizen. The decay of citizenship leads to the decline of the republic and the ascendancy of *fortuna.*"[102] That is, it would mean the end of both republican ideals and the heroic ethic. One of the reasons Machiavelli supported republicanism and the civic militia is because it allowed for the greatest development of *virtu*—in both its senses.

As is often the case, however, Machiavelli's impressive dialectical synthesis did not withstand realpolitik. His work gave birth to the idea of *raison d'état* and so played a key role in the emergence of nationalism. In this case, the republican ideals dropped out but the militarism did not. Friedrich Meinecke explains this phenomenon as follows:

> It has been the fate of Machiavelli, as of so many great thinkers, that only one part of his system of thought has been able to influence historical life. . . . His ideal of *virtu* soon faded . . . and with that too the ethical aim of his statecraft. . . . Generally speaking he was seen first and foremost as having prepared the poison of autocracy; as such, he was publicly condemned and secretly made use of. . . . The chief thing was, however, that the idea of political regeneration was altogether beyond the capabilities and the wishes of the peoples and the rulers of the time, and hence it fell to the ground. . . . Machiavelli's ancient heathen idealism of the State was no longer understood by the men of the Counter-Reformation period. . . . But they very well understood the ancient heathen realism of his statecraft.[103]

Meinecke traces the evolution of Machiavelli's politics as it slowly transformed into a justification for German nationalism in the early twentieth century.[104]

In an attempt to defend Machiavelli against the charges that his theory played into the emergence of nationalism, Maurizio Viroli argues that Machiavelli advocated patriotism rather than nationalism. Viroli defines patriotism as the love of the political institutions, laws, and way of life that sustains the common liberty of the people. Political in orientation, it involves a charitable love of the republic. Nationalism, on the other hand, posits a spiritual unity, a cultural and linguistic oneness or homogeneity among the people. It requires unconditional loyalty and mixes love with pride and fear.[105] But while Viroli rightly distinguishes between patriotism and nationalism, he wishes away the slippage that easily occurs between the two tendencies: Machiavelli theorized patriotism but spawned nationalism. And this is no accident. So while I share Viroli's desire to foster patriotism while condemning nationalism, my understanding of Machiavelli's project reveals that the two go hand in hand. Moreover, when Machiavelli made martial practices the foundation for his republican citizenship, he exacerbated the vicious flip sides of the virtues he was primarily trying to create.

Nevertheless, because of the dialectical nature of his theory, Machiavelli's legacy is appropriately dual. On the one hand his theory did indeed undergo what Meinecke calls a "sinister development" as its dialectical edifice collapsed into "Machiavellism" and then evolved into nationalism[106] — complete with its own virtues and vices. On the other hand, however, Machiavelli's theoretical vision also forms the origins of what Pocock refers to as "the Atlantic republican tradition" that culminated in the American and French Revolutions. This tradition retains Machiavelli's commitment to the cluster of republican ideals: liberty, equality, fraternity, the rule of man-made law, the common good, civic virtue, and participatory citizenship. And we will soon see to what extent the traditional vices continue to live on within this tradition as well. So the dialectical nature of Machiavelli's thought produces two divergent traditions — one more virtuous and one more vicious—each of which entails its own interrelated sets of virtues and vices.

I end with two conclusions and two questions. First of all, based on my reading of Machiavelli it seems that to the extent that the *citizenship of civic practices* privileges martial practices over other possible forms of civic action, the vices of this tradition will be amplified. Due to its militaristic nature, the Citizen-Soldier ideal has two major flaws. In the first place, it requires the presence of an enemy. As Brown puts it, Machiavellian "politics is utterly dependent upon the presence of an enemy, it is at all times a fight, and dissolves when opposition is not present or is too weak to inspire consolidated struggle."[107] The invocation of an enemy for the purposes of fostering republican citizenship brings out the vicious side of the Citizen-Soldier tradition. Constant preparation for combat against an enemy does indeed facilitate the creation of republican citizenship but at an undemocratic price. Preparing for war renders citizen-soldiers patriotic as it fuels their desire for conquest. Citizen-soldiers selflessly serve the republic, but the cause of war exerts pressure on them to conform. Military service necessarily requires both feelings of fraternity and feelings of superiority toward the enemy—chauvinism. So, to the extent that the militaristic practices comprise the civic practices constitutive of citizenship, the vices of civic republicanism will be strengthened. Question number one: Would a broader, less combative variety of civic practices produce the virtues of republican citizenship, while minimizing its related vices?

Secondly, as we have seen in our discussion of Machiavelli, martial practices play a key role in the constitution of both *armed masculinity* and republican citizenship within the Citizen-Soldier tradition; soldiering forms the link that fuses masculinity onto citizenship. Moreover, the combative nature of the *armed masculinity* produced by this tradition results in the denigration of femininity and all the values traditionally associated with it. At the same time, however, because within the *citizenship of civic practices,* both masculinity and citizenship are politically and socially constructed, the possibility remains of reconstructing traditional configuration of gender and citizenship. This leaves us with our second question: What would happen if women began to engage in civic practices that produce masculine citizen-soldiers? In the next chapter we will begin to answer these two questions as we examine the eighteenth century's most important theorist of the Citizen-Soldier ideal: Jean-Jacques Rousseau.

Notes

1. All Machiavelli citations are from *Machiavelli: The Chief Works and Others,* 6 vols., trans. A. Gilbert (Durham, NC: Duke University Press, 1965). For a good recent overview of the debates, see Mark Hulliung, *Citizen Machiavelli* (Princeton: Princeton University Press, 1983), chap. 1. Three good older surveys are Hans Baron, "Machi-

avelli: Republican Citizen and Author of the *Prince," English Historical Review* 76 (1961): 217–53; Richard C. Clark, "Machiavelli: Bibliographic Spectrum," *Review of National Literatures* 1 (1970): 93–135; Eric Cochrane, "Machiavelli: 1940–1960," *Journal of Modern History* 33 (1961): 113–36; and John H. Geerken, "Machiavelli Studies Since 1969," *Journal of History of Ideas* 37 (1976): 351–68.

    2. For Leo Strauss's original argument see his *Thoughts on Machiavelli* (Glencoe, IL: Free Press, 1958). For a defense of Strauss, see Harvey C. Mansfield, Jr., "Strauss's Machiavelli," *Political Theory* 3 (1975): 372–84. For J. G. A. Pocock's criticism see "A Comment on Mansfield's 'Strauss's Machiavelli,' " *Political Theory* 3 (1975): 385–401.

    3. Hulliung, *Citizen Machiavelli.*

    4. Alfred von Martin, *Sociology of the Renaissance* (New York: Harper & Row, 1963), 65–70.

    5. G. W. F. Hegel, *Hegel's Political Writings,* trans. T. M. Knox (Oxford: Clarendon Press, 1964), 219–29. See Shlomo Avineri's commentary in *Hegel's Theory of the Modern State* (Cambridge: Cambridge University Press, 1972), 53–54.

    6. For overviews of these interpretations, see Baron, "Machiavelli," 219; Ernst Cassirer, *The Myth of the State* (New Haven: Yale University Press, 1946), 121–25; and Clark, "Machiavelli," 101. See Friedrich Meinecke's discussion of Hegel, Fichte, and Ranke in *Machiavellism: The Doctrine of Raison d'État and Its Place in Modern History,* trans. Douglas Scott (New Haven: Yale University Press, 1957), 343–91.

    7. For discussions of sixteenth- and seventeenth-century condemnations of Machiavelli and its more favorable eighteenth-century reception, see Baron, "Machiavelli," 217–21; Cassirer, *The Myth of the State,* 116; and Clark, "Machiavelli," 98–101.

    8. For examples, see Jean Bethke Elshtain, *Public Man, Private Woman: Women in Social and Political Thought* (Princeton: Princeton University Press, 1981), 92–99; Max Lerner, introduction to *The Prince and the Discourses* (New York: Random House, 1950); and Harvey C. Mansfield, Jr., *Machiavelli's Virtue* (Chicago: University of Chicago Press, 1996).

    9. Cassirer, *The Myth of the State.*

    10. See Benedetto Croce, *Politics and morals,* trans. Salvatore J. Castiglione (New York: Philosophical Library, 1945), and Maurice Merleau-Ponty, "A Note on Machiavelli," *Signs,* trans. R. C. McCleary (Evanston, IL: Northwestern University Press, 1964), 211–23.

    11. Important republican readings of Machiavelli's work include Sebastian de Grazia, *Machiavelli in Hell* (New York: Vintage Books, 1989); Hannah Fenichel Pitkin, *Fortune Is a Woman: Gender and Politics in the Thought of Niccolo Machiavelli* (Berkeley: University of California Press, 1984); J. G. A. Pocock, *The Machiavellian Moment: Florentine Political Thought and the Atlantic Republican Tradition* (Princeton: Princeton University Press, 1975); and Quentin Skinner, *Machiavelli* (Oxford: Oxford University Press, 1981). For a recent discussion of various aspects of Machiavelli's republicanism, see *Machiavelli and Republicanism,* ed. Gisela Bock, Quentin Skinner, and Maurizio Viroli (Cambridge: Cambridge University Press, 1993).

    12. For example, Baron argues that "instead of looking at the *Prince* and the *Discourses* as two complementary parts of one harmonious whole, we would indeed do better to reconsider what to earlier generations had seemed to be so manifest: that Machiavelli's two major works are in basic aspects different and that the *Discourses* have a message of their own." See "Machiavelli," 217–53. Allan H. Gilbert sees the *Prince* as a

realpolitikal means to a republican end. See his *Machiavelli's Prince and Its Forerunners* (Durham, NC: Duke University Press, 1938) and his introduction to *The Prince and Other Works* (New York: Hendricks House, Farrar, Straus, 1946). Meinecke argues that "the contrast between the monarchist bias in the *Principe* and the republican tinge of the *Discorsi* is only apparent. The quantity of *virtu,* which existed in a people, was the factor that decided whether a monarchy or a republic was the more suitable." See *Machiavellism,* 43.

13. For example, Grazia argues that "a staunch republican, [Machiavelli] is convinced that the times require extraordinary measures taken by one man alone. His republicanism has no theoretical problem accommodating one-generation, one-alone leadership if it will lend life to the republic." See *Machiavelli in Hell,* 240. Meinecke argues that Machiavelli's "republican ideal therefore contained a strain of monarchism, insofar as he believed that even republics could not come into existence without the help of great individual ruling personalities and organizers. He had learnt from Polybius the theory that the fortunes of every State are repeated in a cycle, and that the golden age of a republic is bound to be followed by its decline and fall. And so he saw that, in order to restore the necessary quantum of *virtu* which a republic had lost by sinking to such a low point, and thus raise up the State once again, there was only one means to be adopted; namely, that the creative *virtu* of one individual, of one *mano regia,* one *podesta quasi regia* (*Discourses,* I-18, 55), should take the State in hand and revive it. Indeed he went so far as to believe that for republics which were completely corrupt and no longer capable of regeneration, monarchy was the only possible form of government. Thus his concept of *virtu* formed a close link between republican and monarchical tendencies, and, after the collapse of the Florentine Republic, enabled him without inconsistency to set his hopes on the rule of the Medicis, and to write for them the Book of the Prince. In the same way it made it possible for him immediately afterwards to take up again in the *Discorsi* the strain of republicanism, and to weigh republic and monarchy against one another." See *Machiavellism,* 31–33.

14. Machiavelli, *The Prince,* chap. 26, 95, emphasis mine.

15. Hulliung, *Citizen Machiavelli,* 25.

16. Hulliung, *Citizen Machiavelli,* 19.

17. Hulliung, *Citizen Machiavelli,* 5, emphasis mine.

18. Hulliung begins by arguing that " 'civic humanism,' as formulated by contemporary scholars, errs in de-emphasizing or even expurgating the vital notions of *grandezza* and *gloria* from the republican tradition"—values that he suggests were always there. But he ends by attacking the commitment to republican ideals in Machiavelli's thought: "the political significance of speech and rhetoric has been overemphasized" in republican readings of Machiavelli and so now needs to "suffer a certain demotion in contemporary scholarship." *Citizen Machiavelli,* 21.

19. Pitkin, *Fortune Is a Woman,* 4.

20. Pitkin, *Fortune Is a Woman,* 285.

21. Machiavelli, *Prince,* chap. 12, 47.

22. "War is inescapable. . . . The need for a common defense against other men . . . arises at the dawn of mankind and remains day and night." Grazia, *Machiavelli in Hell,* 166.

23. Pitkin, *Fortune Is a Woman,* 14.

24. See Pocock, *Machiavellian Moment,* and Pitkin, *Fortune Is a Woman,* for a more detailed history.

25. Hans Baron, *The Crisis of the Early Italian Renaissance,* 2d ed. (Princeton: Princeton University Press, 1966).

26. Machiavelli, *Discourses,* III-31, 500.

27. Machiavelli, *Discourses,* II-30, 410.

28. Machiavelli, *Discourses,* II-30, 410–11.

29. For a historical discussion of Machiavelli's practical attempts to organize a militia, see C. C. Bayley, *War and Society in Renaissance Florence* (Toronto: University of Toronto Press, 1961), 240–67.

30. Machiavelli, *Prince,* chap. 12, 48.

31. Machiavelli, *Art of War,* I, 574.

32. Machiavelli, *Prince,* chap. 12, 47. Machiavelli also argues in *Discourses,* I-43 (286) that mercenaries are "useless" because they have "no other reason that holds them firm than the little pay you give them. This reason is not and cannot be enough to make them faithful or so much your friends that they are willing to die for you."

33. Machiavelli, *Discourses,* I-43, 286.

34. Machiavelli, *Discourses,* II-2, 328.

35. Pocock, *Machiavellian Moment,* 200–201.

36. See Pocock, *Machiavellian Moment,* 51–52.

37. Pocock, "A Comment on Mansfield's 'Strauss's Machiavelli,' " 390. See also Grazia, *Machiavelli in Hell,* 216. Cassirer agrees: "In [Machiavelli's] theory all the previous theocratic ideas and ideals are eradicated root and branch." However, despite his rejection of Christianity, Cassirer argues, Machiavelli "never meant . . . to separate politics from religion. He was an opponent of the Church but he was no enemy of religion. He was, on the contrary, convinced that religion is one of the necessary elements of man's social life. But in his system this element cannot claim any absolute, independent, and dogmatic truth. Its worth and validity depend entirely on its influence on political life. By this standard, however, Christianity occupies the lowest place. For it is in strict opposition to all real political *virtu.* . . . A merely passive religion, a religion that flees the world instead of organizing it, has proved to be the ruin of many kingdoms and states." *The Myth of the State,* 138.

38. Pocock, *Machiavellian Moment,* 157, 193.

39. For an oppositional discussion of virtue in Machiavelli, see Mansfield, *Machiavelli's Virtue,* 6–52. For example, Mansfield asserts the "classical republican interpretation[s] . . . understand Machiavelli's virtue admiringly as self-sacrifice for the common good of the republic. That it is not" (xv).

40. Quentin Skinner, "The Republican Ideal of Political Liberty," in Bock, Skinner, and Viroli, *Machiavelli and Republicanism,* 303, emphasis mine.

41. Grazia, *Machiavelli in Hell,* 176.

42. Skinner, "The Republican Ideal of Political Liberty," 303.

43. Pocock, *Machiavellian Moment,* 201.

44. Pocock, *Machiavellian Moment,* 203.

45. Most scholars agree with this derivation of the term *virtu.* For an oppositional view, see Mansfield, who denies that Machiavelli's virtue comes from ancient or Roman understandings of manliness. *Machiavelli's Virtue,* 36–37.

46. Pitkin, *Fortune Is a Woman,* 25. For similar arguments see Hulliung, *Citizen Machiavelli,* 29, and Wendy Brown, *Manhood and Politics: A Feminist Reading in Political Theory* (Totowa, NJ: Rowman & Littlefield, 1988), 90.

47. For an in-depth and nuanced exploration of this issue, see Pitkin, *Fortune Is a Woman.*
48. Machiavelli, *Prince,* chap. 25, 92.
49. Machiavelli, *Discourses,* II-2, 330–31, emphases mine.
50. Hulliung, *Citizen Machiavelli,* 68.
51. Bonnie Honig, *Political Theory and the Displacement of Politics* (Ithaca: Cornell University Press, 1993), 68–69.
52. Honig, *Political Theory and the Displacement of Politics,* 230.
53. Honig, *Political Theory and the Displacement of Politics,* 3.
54. Pitkin, *Fortune Is a Woman,* 117–18.
55. Brown makes a similar argument in *Manhood and Politics.*
56. Pitkin, *Fortune Is a Woman,* 136, emphasis mine.
57. Brown, *Manhood and Politics,* 88.
58. Machiavelli, *Discourses,* I-21, 246, emphasis mine.
59. Machiavelli, *Discourses,* I-21, 247, emphasis mine.
60. Machiavelli, *Art of War,* I, 581, emphasis mine.
61. Machiavelli, *Prince,* chap. 26, 94.
62. Machiavelli, *Discourses,* I-4, 202.
63. Machiavelli, *Discourses,* I-34, 267.
64. Machiavelli, *Discourses,* I-3, 201.
65. Machiavelli, *Discourses,* I-3, 201, emphasis mine.
66. Machiavelli, *Discourses,* I-35, 270.
67. Machiavelli, *Discourses,* I-34, 268.
68. Machiavelli, *Discourses,* III-22, 482.
69. Machiavelli, *Discourses,* II-2, 329.
70. Machiavelli, *Discourses,* I-49, 296.
71. See Pocock, *Machiavellian Moment,* 193.
72. Machiavelli, *Discourses,* I-29, 257.
73. Machiavelli, *Discourses,* II-1, 323.
74. Machiavelli, *Discourses,* I-58, 315.
75. Grazia, *Machiavelli in Hell,* 269–70.
76. Brown, *Manhood and Politics,* 73.
77. Machiavelli, *Discourses,* I-16, 235, emphasis mine. Also see Brown, *Manhood and Politics,* 74.
78. Brown, *Manhood and Politics.*
79. Elshtain, *Public Man, Private Woman,* 98.
80. Machiavelli, *Discourses,* I-21, 247; III-36, 510.
81. Machiavelli, *Discourses,* I-18, 242.
82. Machiavelli, *Discourses,* I-58, 316.
83. Machiavelli, *Discourses,* III-9, 453.
84. Machiavelli, *Discourses,* I-3, 203.
85. Machiavelli, *Discourses,* I-3, 202.
86. Machiavelli, *Discourses,* I-4, 203.
87. Pitkin stresses this point in *Fortune Is a Woman,* chap. 4.
88. Grazia, *Machiavelli in Hell,* 192.
89. Pitkin, *Fortune Is a Woman,* 93.
90. Pitkin, *Fortune Is a Woman,* 96.

91. Machiavelli, *Discourses,* III-28, 492.
92. Machiavelli, *Discourses,* III-28, 493.
93. Honig, *Political Theory and the Displacement of Politics,* 70. See Machiavelli, *Discourses,* 2. Preface; I.vi. and "Homer's Contest," 36–37.
94. Machiavelli, *Discourses,* II-25, 399.
95. Machiavelli, *Discourses,* III-16, 469.
96. Honig, *Political Theory and the Displacement of Politics,* 71. See also Hulliung, *Citizen Machiavelli,* 26.
97. Machiavelli, *Discourses,* I-6, 211.
98. Machiavelli, *Discourses,* III-31, 500, emphasis mine.
99. Machiavelli, *Discourses,* I, 578.
100. Hulliung, *Citizen Machiavelli,* 36; Grazia, *Machiavelli in Hell,* 172.
101. Hulliung, *Citizen Machiavelli,* 5–6.
102. Pocock, *Machiavellian Moment,* 56. Hulliung argues similarly: "While the individual excellence of the prince may be admirable, the greatest feats of heroism are collective and popular in nature. In its democratic form, *virtus* taps the potential greatness of the common man, his willingness to fight and die for his country, and can claim as its due meed of glory the conquest of all other republics." *Citizen Machiavelli,* 5–6.
103. Meinecke, *Machiavellism,* 44–45.
104. See also Cassirer, *The Myth of the State,* 140–41.
105. Maurizio Viroli, "The Meaning of Patriotism," paper presented at the Walt Whitman Seminar, Rutgers University, New Brunswick, NJ, 1 February 1994.
106. Meinecke, *Machiavellism,* 410.
107. Brown, *Manhood and Politics,* 115.

Chapter 3

"Jean-Jacques . . . You Are a Genevan": Civic Festivals, Martial Practices, and the Production of Civic Identity

Your true republican is a man who . . . has eyes only for the fatherland.

—Rousseau, *The Government of Poland*

Rousseau's political theory presents us with the best-articulated eighteenth-century version of the Citizen-Soldier tradition. Although Rousseau's Citizen-Soldier tradition parallels Machiavelli's to some extent, it also differs from Machiavelli's in several interesting ways. This chapter argues that the Citizen-Soldier stands at the very center of Rousseau's theoretical framework because it embodies a set of practices that produce the necessary foundation for republican self-rule: The civic and martial practices constitutive of the Citizen-Soldier also produce patriotism, fraternity, and civic virtue, as well as the *general will* itself, the essential prerequisites for government aimed at the common good. At the same time, however, the practices of the Citizen-Soldier tradition also produce a set of vices that form the flip sides of those same virtues: patriotism can produce nationalism, fraternity can produce fusion, civic virtue can produce homogeneity, and the *general will* can produce a totalizing civic identity.

While these interrelated virtues and vices are always directly related, Rousseau's version of the Citizen-Soldier actually exacerbates the vicious side of this tradition because he creates unity through an all-encompassing set of civic and martial practices, the channeling of all passion toward the fatherland, and the production of a totalizing civic identity that replaces all others. That is to say, while he avoids the problems associated with the creation of unity through opposition to an external enemy—Machiavelli's vice—Rousseau creates unity by forging versions of fraternity, patriotism, and civic identity that are so strong that they slip easily into fusion, nationalism, and homogeneity. Nevertheless, as with Machiavelli, Rousseau's political legacy is appropriately dual. His theory of the Citizen-Soldier contributed both to the radical democratic concept of popular sovereignty that underwrote the "Age of Democratic Revolution" and to the emergence of nationalism in post-Revolutionary Europe. The chapter concludes with an examination of the transgressive struggle of *La Societe des Citoyennes Republicaines Revolutionaires* to occupy the position of the Citizen-Soldier.

Rousseau and Machiavelli

At first glance, Rousseau's argument for the civic militia parallels Machiavelli's in several ways. Like Machiavelli, Rousseau argues that a civic militia made up of all citizens forms the best mode of defense for a republic because soldiers fight most effectively when defending their own liberty: "people always fight better in defense of their own than in defense of what belongs to others."[1] Consequently, citizens make the best soldiers. Secondly, like Machiavelli, Rousseau emphasizes the importance of a civic militia as a bulwark against tyranny. And while Rousseau, writing in the eighteenth century, feared a tyranny imposed by the king's standing army rather than by mercenaries, his point is the same:

> Regular armies have been the scourge and ruin of Europe. They are good for only two things: attacking and conquering neighbors, and fettering and enslaving citizens. . . . The state's true defenders are its individual citizens, no one of whom should be a professional soldier, but each of whom should serve as a soldier as duty requires.[2]

The protection of the republic from tyranny requires "a good militia"—"a genuine, well-trained militia"—to "be ready at all times to serve the republic."[3] And thirdly, like Machiavelli, Rousseau says that only defensive military forces are needed in a republic. Because a republic is concerned fundamentally with protecting liberty, "offensive power is incompatible with [a republican] form of government. Those who will freedom must not will conquest as well."[4] Furthermore, "he who tries to take away the freedom of others nearly always ends up losing his own."[5] According to this traditional argument, the civic militia is the best defense against both tyranny and war.

And like Machiavelli—or any theorist of the Citizen-Soldier tradition, for that matter—Rousseau presents a model of citizenship that I have named the *citizenship of civic practices*. According to this paradigm, "citizens" are not pre-existing entities who then choose whether or not to engage in political action. Instead, citizens are actually *produced* through engagement in civic and martial practices. One is not born but rather becomes a citizen as one engages with others in civic practices, in particular those of the civic militia. And while the Citizen-Soldier tradition requires both participation in legislation for the common good and service in the militia, it is the latter that forms the necessary prerequisite for the willingness to do the former.

As with Machiavelli, Rousseau's Citizen-Soldier constitutes a normative ideal that links military service to participatory citizenship. Citizen-soldiers fight to defend not only their right to self-govern and but also the corresponding republican ideals of liberty, equality, fraternity, the rule of law, the common good, civic virtue, and participatory citizenship. Because of this connection, both halves of the Citizen-Soldier are equally important. In other words, the Citizen-Soldier ideal cannot be equated simply with universal conscription. Instead, it links universal conscription with participatory citizenship.

Finally, Rousseau's Citizen-Soldier ideal, like Machiavelli's, fuses *armed masculinity* onto republican citizenship. For Rousseau, being a man means being a soldier; the two identities are mutually constituted through engagement in martial practices. The practices dictated by the Citizen-Soldier ideal fuse this performatively constructed *armed masculinity* onto republican citizenship. As with Machiavelli, Rousseau's male individuals become manly republican citizen-soldiers as they engage in the civic and martial practices prescribed by the ideal of the Citizen-Soldier.

Despite these similarities, however, Rousseau's version of the Citizen-Soldier tradition differs significantly from Machiavelli's. Rousseau's *citizenship of civic practices* includes a much broader array of civic practices than does Machiavelli's. While Machiavelli focuses almost exclusively on martial practices as creating the necessary prerequisites to self-government aimed at the common good, Rousseau sees citizenship as constituted not only through martial practices but also through the wide variety of civic practices comprising the civic festival. Martial practices play a central but not an exclusive role in Rousseau's vision.

Machiavelli and Rousseau create civic unity in profoundly different ways and so present us with different dangers. Machiavelli imagines a unitary citizenry formed in opposition to an external enemy, and this emphasis ends up leading Machiavelli away from his claim about the defensive nature of the civic militia and toward the valorization of conquest. Rousseau, on the other hand, resists using the threat of an external enemy to unify his citizenry. While he does make some use of the existence of *internal* enemies—in the guise of femininity and particularity—for the most part he creates civic unity by envisioning an all-encompassing set of civic and martial practices that produce not only a totalizing civic identity that leaves no room for particularity, but also versions of civic virtue, fraternity, and patriotism so strong that they easily slip over into homogeneity, fusion, and nationalism.

Rousseau's *Citizenship of Civic Practices*

In keeping with the Citizen-Soldier tradition, Rousseau presents us with a *citizenship of civic practices* in which individuals *become* citizens only as they engage together in civic and martial practices. For example, in *On the Social*

Contract Rousseau defines citizenship as civic participation: "As for the associates, they collectively take the name *people;* and *individually are called Citizens as participants in the sovereign authority,* and *Subjects* as subject to the laws of the State."[6] Individuals are only citizens to the extent that they participate in sovereign authority, that is, in the civic practices of self-government. In other words, "citizens" are not pre-existing entities who then decide whether to engage in politics or not. Instead, the category of "citizen" is constituted only through engagement in civic and martial practices. Put differently, the founding of a republic creates the possibility of citizenship because it provides individuals with the opportunity to become citizens as they *participate* in the political process.

Rousseau's social contract reconciles the individual and the community. Although the social contract transforms "private person[s]" into a "moral and collective body" of citizens, these citizens do not lose their standing as individuals within the republic: "In addition to the public person, we have to consider the private persons who compose it and whose life and freedom are naturally independent of it."[7] In addition to maintaining "life and freedom" outside of the body politic, the individual also retains his own individual voice within the republic. As Rousseau puts it, the "moral and collective body" created by the social contract is "composed of as many *members* as there are *voices* in the assembly."[8] Rousseau attempts to reconcile the individual with the community; one does not stand opposed to the other.

Participatory citizenship is essential to Rousseau's vision. The people must take an active role in governing themselves: "If the people promises simply to obey, it dissolves itself by that act; it loses the status of a people."[9] Citizenship—and even the concept of "the people"—requires civic participation. Unlike Kant's *categorical imperative,* Rousseau's *general will* cannot be determined in the abstract.[10] It requires hard political work. The "pluses and minuses" of particular wills "which cancel each other out" can only be known in practice.[11] No one individual alone can do the work required of citizens. And moreover, outside of the context of political participation, there are no citizens, because citizenship must be constantly constructed and reconstructed through engagement in civic practices.

Moreover, participation in civic practices actually produces a new civic identity that replaces other, more particularistic identities.[12] That is, because identity is performatively constructed, as the individual participates in civic republican practices, he undergoes a change of identity. "*In place of* the private person" stands the citizen, an identity "produce[d]" through civic practices, such as participation in political assemblies and service in the civic militia.[13] And for Rousseau, particular identities are not simply supplemented by the new civic identity, but instead are *replaced.* Rousseau's vision has no room for multiple identities. One cannot be, for example, both a citizen and a Roman Catholic, because such split identities would divide the republic and "everything that destroys social unity is worthless."[14] Or as Rousseau remarks in the *Government of Poland,* the

Polish citizen should be "purely Polish" and nothing else. The new civic identity does not synthesize a multiplicity of particular identities. Instead, it replaces them.

Particular interests as well as particular identities must be discarded in order to discover the *general will*. That is, the *general will* by definition includes only what citizens share in common: "There is often a great difference between the will of all and the general will. The latter considers *only the common interest;* the former considers private interest, and is only a sum of private wills."[15] The *general will* concerns what citizens share in common and excludes what differentiates them from one another.

Essentially, what Rousseau does is create unity out of diversity, albeit by focusing only on what individuals share in common. That is to say, Rousseau begins with the reality of the diversity of individuals—not diverse by today's standards, granted, but diverse in the sense that each individual has his own unique desires, interests, and beliefs. Put differently, the political process as articulated by Rousseau allows for a certain degree of heterogeneity, in that he assumes there will be a multiplicity of conflicting "private wills" among citizens. As he puts it, men "may be *unequal* in force or in genius."[16] Differences of wealth and rank may continue to exist within a republic without being incompatible with political equality. The important point is that these differences should not affect citizenship, in that all men, regardless of particularities, can and should be transformed into citizens through engagement in civic practices, such as soldiering. Precisely because a certain degree of heterogeneity necessarily exists within any republic, engaging together in civic practices is essential to the constitution of republican citizenship.

Again the critical point here is that citizenship must be a relationship of *political* equality in the face of individual differences. For Rousseau, the formation of the social contract actually *creates* political equality: "The fundamental compact . . . substitutes a moral and legitimate equality for whatever physical *inequality* nature may have placed between men, and that although they may be *unequal* in force or in genius, they all become equal through convention and by right."[17] The rule of law within the community, self-given law, ensures the political equality of all citizens:

> The social pact establishes equality among the citizens in that they all pledge themselves under the same conditions and must all enjoy the same rights. Hence by the nature of the compact, every act of sovereignty, that is, every authentic act of the general will, binds or favours all the citizens equally, so that the sovereign recognizes only the whole body of the nation and makes no distinction between any of the members who compose it.[18]

Republican citizenship necessarily entails political equality.

Nevertheless, as Iris Marion Young and Joan Landes rightly argue, Rousseau's political equality and *general will* come at the expense of particularity and difference. They criticize Rousseau's version of the common good for excluding from political deliberations everything not held in common.[19] In other words, Rousseau's *general will* does not allow room for serious dissension and debate. As Rousseau explains:

> As long as several men together consider themselves to be a single body, they have *only a single will,* which relates to their common preservation and the general welfare. Then all the mechanisms of the state are vigorous and simple, its maxims are clear and luminous, it has *no tangled, contradictory interests;* the common good is *clearly apparent* everywhere, and requires only good sense to be perceived. . . . But when the social tie begins to slacken and the State to grow weak; when private interests start to make themselves felt and small societies to influence the large one, the common interest changes and is faced with opponents; unanimity no longer prevails in the votes; the general will is no longer the will of all; *contradictions and debates arise* and the best advice is *not accepted without disputes.*[20]

Rousseau does not imagine a *general will* that is created out of real political deliberation among diverse individuals with (initially) very different agendas. Instead, his *general will* requires the elimination of particularity from the political discussion. Only with this done can citizens focus on what they share in common. Rousseau may create unity out of diversity, but only at the expense of diversity.

As Young and Landes argue, because the *general will* "considers only the common interest," it cannot address the particular issues that concern only certain sectors of the population. So, for instance, Rousseau's political process cannot include the so-called "women's issues" because these issues are seen as particular rather than universal. Thus, Young and Landes conclude, the usefulness of Rousseau's concept of the *general will* for contemporary American politics is limited. And on this point they are right: A conception of the common good that cannot accommodate real debate and that on principle cannot address political issues of concern only to particular sectors of the citizenry, such as women, African Americans, lesbians, or gay men, should not be incorporated into contemporary American politics. Nevertheless, despite the limitations of Rousseau's *general will,* this does not mean, I would argue, that every rendition of the common good has to be exclusionary, as Young and Landes imply. However, as with other aspects of Rousseau's theoretical vision, his *general will* overemphasizes unity and so ends up becoming exclusionary and thus undemocratic—a point we will address more fully later.

These problems notwithstanding, however, Rousseau's restriction of the *general will* to what citizens share in common prevents his political theory from becoming totalitarian. The *general will* itself is not a totalitarian concept. As Rousseau emphasizes, the *general will* does allow room for personal freedoms, for the pursuit of "particular desires, needs, and interests" and for "personal satisfaction" but only *outside of the body politic;* the individual maintains "life and freedom . . . *independent of*" the body politic. That is, "each person alienates through the social compact only that part of his power, goods, and freedom whose use matters to the community." He does not abandon everything. And while "the sovereign alone is the judge of what matters, . . . the sovereign, for its part, cannot impose on the subjects any burden that is useless to the community."[21] And of course, we must remember here that for Rousseau the sovereign is nothing other than the body politic made up of all citizens. Consequently, Rousseau can plead innocent to the charges of "totalitarianism" as leveled by J. L. Talmon.[22]

Civic Festivals: Creating the Foundation for the *General Will*

Rousseau begins *On the Social Contract* by puzzling over how to reconcile individual freedom with civil society and the rule of law:

> "Find a form of association that defends and protects the person and goods of each associate with all the common force, and by means of which each one, uniting with all, nevertheless obeys only himself and remains as free as before." This is the fundamental problem which is solved by the social contract.[23]

Rousseau's reconciliation of individual freedom with a civil society governed by the rule of law depends upon his definition of freedom. Rousseau defines freedom, not as the absence of law, but rather as obedience only to one's own will: When an individual obeys only himself, he is free. Working from this definition, Rousseau reasons that if freedom is obedience only to one's own will, then an individual remains free when he obeys self-willed law: "Obedience to the law one has prescribed for oneself is freedom."[24] In this way, Rousseau reconciles freedom with obedience to the law.

This solution leads Rousseau to another puzzle, however: What guarantees that a man will obey the law he gives to himself? As Rousseau phrases this question (in a different text): "By what means are we to move men's hearts and bring them to love their fatherland and its laws?"[25] The cursory answer he presents in *On the Social Contract* is that "mores, customs, and especially . . . opinion—a part of the laws unknown to our political theorists, but on which the success or all the others depends"—play a key role in creating the foundation for republican self-rule. This "fourth" and "most important" type of law—mores, customs, and opinion—is

> not engraved on marble or bronze, but in the hearts of the citizens; which is the true constitution of the state; which gains fresh force each day; which, when other laws age or die out, revives or replaces them, preserves a people in the spirit of its institution, and imperceptibly substitutes the force of habit for that of authority.[26]

In other words, the existence of common mores, customs, and opinions forms the foundation for the existence of the *general will* and for the willingness to obey its laws.

The Citizen-Soldier stands at the very center of Rousseau's entire theoretical framework because this figure embodies the practices that produce the necessary foundation for the existence of a *general will* and obedience to the laws it generates. More specifically, the willingness of the citizen to obey self-given laws depends upon the existence of patriotism, fraternity, and civic virtue—all of which are created through engagement in civic and martial practices. Patriotism, fraternity, and civic virtue are not natural attributes of men. Instead, they must be created within the right institutional context. The "spiritual vigor" and "patriotic zeal" characteristic of the ancients are "unknown to us moderns," Rousseau tells us, only because we do not use our political institutions to construct these qualities as a second nature. "What prevents us from being the kind of men [the ancients] were?" Rousseau asks. His answer: "The prejudices, the base philosophy, and the passions of narrow self-interest which, along with indifference to the welfare of others, *have been inculcated in all our hearts by ill-devised institutions.*"[27] The institutional context is the primary determinant of how men will behave. Only engaging together in civic practices will transform self-interested individuals into republican citizens, who have a *general will* and will consider it rather than just their own, narrowly defined interests. For this reason, institutionalized arenas for civic action are critical to Rousseau's vision: Without them we will have not citizens obedient to the *general will* but, rather, self-interested individuals.

Civic festivals play an important role in the creation of patriotism, which forms a necessary precondition for obedience to the law. "By what means," Rousseau asks, "are we to move men's hearts and bring them to love their fatherland and its laws? Dare I say? Through the games they play as children, through institutions that, though superficial man would deem them pointless, develop habits that abide and attachments that nothing can dissolve."[28] Civic-oriented games teach children patriotism by teaching them to love and respect the principles of civic republicanism. Playing together in public gets children "accustomed, from an early moment, to rules, to equality, to fraternity, to competitions, to living with the eyes of their fellow citizens upon them, and to seeking public approbation."[29] In this way, it prepares them for republican citizenship. Participation in civic practices—in public play as children and in the civic militia as men—creates patriotism.

Civic festivals also play a key role in the constitution of republican citizens because they help create fraternal bonds among individuals, which form a necessary prerequisite for the *general will*. "These festivals," Rousseau tells us, "serve many useful purposes which would make of them an important component of the training in law and order and good morals [manners]."[30] At the civic festivals, Rousseau argues, "each sees and loves himself in the others so that all will be better united."[31] Its purposes include "bring[ing] people together not so much for a public entertainment as for the gathering of a big family, and *from the bosom of joy and pleasures would be born the preservation, the concord, and the prosperity of the republic."*[32] Participating in civic festivities together creates a sense of fraternity that underlies the existence of the *general will.*

Moreover, civic festivals contribute to the constitution of a common civic identity among diverse individuals. As Mona Ozouf argues, one of the "great mythical experiences" of the eighteenth century was "the individual who is re-baptized as citizen in the festival." In the Enlightenment world of the eighteenth century, "men were individuals, in theory all identical, all equal, but solitary." However, "through the festival [a] new social bond was to be made manifest, eternal, and untouchable." Echoing Rousseau, she explains that "the festival was an indispensable complement to the legislative system, for although the legislator makes the laws for the people, festivals make the people for the laws."[33] Individuals could become citizens—become a people—only as they participated together in civic practices, such as those comprising the festivals.

In fact, civic festivals played a key role in the construction of Rousseau's own civic identity. In his famous footnote in *The Letter to M. d'Alembert,* Rousseau recounts an early experience with a civic festival as a formative incident in the creation of his own civic identity: "I remember having been struck in my child-hood by a rather simple entertainment, the impression of which has nevertheless always stayed with me in spite of time and variety of experience." Rousseau's father explicitly spelled out the meaning of the civic festival to the young man:

> "Jean-Jacques," he said to me, "*love your country.* Do you see these good Genevans? They are all *friends,* they are all *brothers;* joy and concord reign in their midst. *You are a Genevan;* one day you will see other peoples; but even if you should travel as much as your father, you will not find their likes."[34]

Clearly, for the young Rousseau, witnessing the civic festivities of his community played an essential role in the constitution of his civic identity. He essentially became a Genevan as he participated in civic festivals with other Genevans.

The practices of the civic militia form the centerpiece of the civic festival as both described and prescribed by Rousseau:

> The regiment of Saint Gervais had done its exercises, and, according to the custom, they had supped by companies; most of those who formed them gathered after supper in the St. Gervais square and started dancing all together, officers and soldiers, around the fountain, to the basin of which the drummers, the fifers and the torch bearers had mounted.[35]

The spectacle of these martial practices instilled in the young Rousseau a passionate allegiance to Geneva:

> A dance of men, cheered by a long meal, would seem to present nothing very interesting to see; however, the harmony of five or six hundred men in uniform,

holding one another by the hand and forming a long ribbon which wound around, serpentlike, in cadence and without confusion, with countless turns and returns, countless sorts of figured evolutions, the excellence of the tunes which animated them, the sound of the drums, the glare of the torches, a certain military pomp in the midst of pleasure, *all this created a very lively sensation that could not be experienced coldly.* . . . There resulted from all this *a general emotion* that I could not describe but which, in universal gaiety, is *quite naturally felt in the midst of all that is dear to us.*[36]

As this passage demonstrates, participation in civic festivities produces a civic identity, feelings of fraternity, and a passionate patriotism, all of which undergird civic participation, military service, and the *general will* itself.

The martial practices that form the centerpiece of the Rousseauian civic festival also play a key role in the creation of civic virtue. Engagement in martial practices creates the "strength and vigor" of the body necessary for both the ability and the willingness to defend republican liberty: "What view of hunger, thirst, fatigues, dangers, and death can men have if they are crushed by the smallest need and rebuffed by the least difficulty? Where will soldiers find the courage to bear excessive work to which they are totally unaccustomed?"[37] Rousseau argues that the aristocratic armies of his day pale in comparison to the more vigorous ancients:

If the strength of the men of antiquity is compared to that of the men of today, no sort of equality can be found. Our gentlemen's exercises are children's games next to those of ancient gymnastic; rackets (*la paume*) has been abandoned as too fatiguing, and we can no longer travel by horseback. I say nothing of our troops. The marches of the Greek and Roman armies can no longer be conceived. Just to read of the length of march, the work, and the burden of the Roman soldier is tiring and overwhelms the imagination. . . . We are fallen in everything.[38]

Rigorous martial practices are necessary to produce soldiers physically capable of defending republican liberty.

According to Rousseau, the creation of a vigorous and healthy body forms the necessary prerequisite for the existence of civic virtue—the willingness to obey the *general will*. Rejecting the Cartesian mind/body split, Rousseau sees a direct connection between the will of the soul—a combination of heart and mind—and the vigor of the body. Martial practices are essential to the production of citizens because they provide the foundation for the existence of virtue—defined as "strength and vigor of the soul"—which is the necessary prerequisite for government in accordance with the *general will*. Martial practices do this by providing virtue's necessary foundation: "strength and vigor of the body."[39] "Military exercise" makes men virtuous by making them "vigorous and warlike" in both body and soul.[40] That is why, "the ancient Greek republics . . . forbade their citizens the practice of those tranquil and sedentary occupations which, by weighing down and corrupting the body, soon enervate the vigor of the soul."[41] When in Rome "military discipline was neglected," Rousseau tells us, "virtue . . . was lost."[42] A weak body falls prey to the appetites. Only a robust body can place virtue over desire and particularity.

Rousseau valorizes the ancients—particularly, the Spartans and the Roman republicans—because they understood the ways in which engagement in martial practices contributed to the constitution of citizens. More specifically, the ancients used their military institutions to create the virtues that are the necessary foundation of patriotism and fraternity—all necessary characteristics of republican citizenship. That is to say, Rousseau argues that military exercises produce the "moral qualities" necessary to patriotism: They create "magnanimity, equity, temperance, humanity, [and] courage" without which "that sweet name fatherland will never strike [man's] ear," will be "forgotten," or will be at least "disdainfully" dismissed.[43] And without a sense of patriotism, "unity" will disintegrate and "sects" will be "embraced."[44] Consequently, the republic will cease to exist, and we will "no longer have citizens."[45] Thus, Rousseau praised the ancients because their military practices created the moral virtues, patriotism, and fraternity—all qualities necessary for the republic and its citizen-soldiers.

Thus, the ideal of the Citizen-Soldier stands at the center of Rousseau's entire theoretical vision. This figure represents the integration of a variety of civic practices, all of which are essential for republican citizenship. Both halves of the Citizen-Soldier are equally important. Citizen-soldiers acquire a common civic identity and learn patriotism, fraternity, and civic virtue through engagement in civic practices, including but not limited to military service. And it is the acquisition of these characteristics which undergirds the willingness of citizen-soldiers both to govern themselves in accordance with the *general will* and to risk their lives to defend the republic through participation in the civic militia. Citizen-soldiers make laws for themselves and are willing to defend their ability to do so.

"In a Republic, Men Are Needed"

The figure of the Citizen-Soldier fuses *armed masculinity* onto republican citizenship. That is to say, the practices productive of citizen-soldiering within the civic republican tradition also construct *armed masculinity*. For example, soldiering is central both to what it means to be a citizen and to what it means to be a man. We can see the conflation of the three identities—man, soldier, and citizen—in the following passage:

> In Switzerland, every bridegroom must have his uniform (it forthwith becomes the suit he wears on feast-days), his regulation rifle, and the full equipment of a foot

soldier. He is at once enrolled in the local company of the militia. In the summer-
time, on Sundays and on feast-days, he and his fellows are put through drills in accordance with the schedules for the several rosters.[46]

In order to participate in the institution of marriage, an essential rite of passage
into manhood, a male individual must wear a military uniform and bear arms:
He must *become* a soldier. Secondly, becoming a soldier requires active participation in the civic militia, which links soldiering to citizenship. And finally,
the manly citizen-soldier must wear his uniform and participate in martial
"drills" as part of the civic festivities of his community held "in the summer-
time, on Sundays and on feast-days." To be a man, he must be a soldier, and to
be a soldier, he must also be a citizen. In sum, soldiering is central both to what
it means to be a man and to what it means to be a citizen. Man, soldier, and citizen are one and the same. Thus, Rousseau presents participation in the civic
militia as central to masculinity, to citizenship, and to civic life in general.

In arguing that the same practices produce citizenship and masculinity simultaneously, my analysis differs from Judith Shklar's classic republican interpretation of Rousseau's work, *Men & Citizens.*[47] In that work, Shklar argues that
Rousseau presents two competing and mutually exclusive models of society, one
for men and one for citizens: "One model was a Spartan city, the other a tranquil
household, and the two were meant to stand in polar opposition to each other."[48]
Men have to choose one or the other ideal, she argues; they cannot use both.[49]
Texts such as the *New Heloise* and the *Emile* present a model of society for men,
Shklar argues, while the *Discourse on the Arts and Sciences, The Government of
Poland,* and the *Social Contract* present a political society for citizens. In opposition to that argument, I have shown that the texts Shklar designates as examples
of the citizen model are as much about manhood as they are about citizenship.

Citizenship and masculinity are profoundly interconnected for Rousseau because both identities are performatively constructed through the same set of civic
and martial practices. And because they are performatively constructed, they can
never be permanently achieved. As Linda Zerilli demonstrates, for Rousseau
"sexual desire and gender difference emerge in society; they are not natural facts
but *performative enactments.*"[50] Likewise, Elizabeth Wingrove argues that
Rousseau made the "realization that gender is an assumed identity. . . . *Becoming*
a man *means* correctly *performing* a role." Consequently, for Rousseau "the on-
tological and psychic status of 'maleness' becomes as problematic as that of the
citizen. [Thus,] it would seem that both the man and the citizen need to be produced"[51]—and produced simultaneously. In other words, because of their performative nature, both masculinity and republican citizenship must be continually
constructed and reconstructed through engagement in civic and martial practices.

Rousseau sees martial practices as essential to the construction of *armed
masculinity,* which is central to his version of republican citizenship. As he puts
it, "the first Romans *lived like men* and found in their *constant* [military] exer-

cises *the vigor that nature had refused them.*"[52] Not masculine by *nature,* the Romans were men only because they "lived like men." Moreover, because one is not born but rather *becomes* a man—to expand upon Simone de Beauvoir's famous insight (and Judith Butler's interpretation of it)—the acquisition of masculinity is a process that can never be finally finished.[53] As Rousseau says, the Romans became masculine only through *"constant* exercise." Only continual engagement in martial practices prevents the emergence of a "softness of character"[54]—of femininity—that always threatens to disrupt the fragile construction of *armed masculinity.* For example, Rousseau condemns his contemporary males for losing their masculine "vigor . . . in the indolent and soft life to which our dependence on women reduces us." Because gender is performatively constructed, males become *like* women when they *"live like them."*[55] To prevent this dissolution of masculinity, Rousseau advocates military exercises to teach participants "what they *ought to do as men.*"[56] He wants "military discipline" to produce the manly "military virtues." Otherwise, "effeminate customs" will replace "heroic actions."[57] Rousseau wants to remember the important role of martial practices in the production of manly citizen-soldiers.

Since masculinity is always constructed in opposition to femininity, "women" must be excluded from the practices constitutive of masculinity. Women do not participate in the martial practices that are central to republican citizenship. Nor can they join in the civic deliberations of the male-only *circles.* At civic festivals, married women serve as spectators only. And while girls do participate in the civic festivals, the lessons they learn there differ markedly from those of the boys. In Wingrove's words, "it is around and through sexual roles that the festival itself is organized. . . . Its central focus is to make [people] *un*equal as it makes them men and women, as it makes them ruler and ruled."[58] Whereas boys participate in civic festivals as a preparation for citizenship, girls participate in order to meet a husband. Marriage for boys forms the gateway into civic life. For girls, it marks a final exit.

Since gender identity is performatively constructed—engagement in "masculine" practices produces "men" and engagement in "feminine" practices produces "women"—Rousseau envisions very different roles for boys and girls who participate in civic festivals. For example, Rousseau wishes "that every year, at the last ball, the young girl, who during the preceding one has comported herself most decently, most modestly, and has most pleased everyone . . . be honored with a crown . . . and the title of Queen of the Ball."[59] Participation in the feminine version of civic practices during the festivals teaches young girls decency, modesty, and charm. Ultimately, femininity will require the continual exclusion from civic practices.

At the same time, however, engagement in masculine civic practices prepares young boys to become masculine citizen-soldiers. For example, Rousseau wants contemporary boys to be "rustically raised," as they were in his time. During that utopian era,

the fathers took the [boys] with them on the hunt, in the country, to all their exercises, in every society. Timid and modest before aged people, [the boys] were
hardy, proud, and quarrelsome among themselves. They had no hairdo to preserve;
they challenged one another at wrestling, running, and boxing. They fought in
good earnest, hurt one another sometimes, and then embraced in their tears. They
went home sweating, out of breath, and with their clothes torn; they were real
scamps, *but these scamps made men who have zeal for the service of the country
in their hearts and blood to spill for it.*[60]

The games boys play as children prepare them for participation in the civic and
martial practices constitutive of masculine citizen-soldiering. On the other
hand, the games girls play as children prepare them for the exclusion from civic
and martial practices that constitutes femininity.

While Rousseau's vision requires the participation of women as spectators
in civic festivals, they must not be included in the practices constitutive of
armed masculinity and republican citizenship. So during the idealized martial
festival of Rousseau's youth, when "five or six hundred men in uniform, [held]
one another by the hand and form[ed] a long ribbon which wound around, serpentlike, in cadence and without confusion," at least as Rousseau tells it, "the
women were in bed." Undisturbed, the citizen-soldiers of Rousseau's republic
exuberantly engaged in homosocial martial practices that constituted them as
masculine citizen-soldiers.

As soon as they realized the festival had begun, however, the women "came
down" to the public square—perhaps wishing to engage in the civic practices
that would constitute them as republican citizens alongside men. Nevertheless,
the arrival of the women disrupted the dance of republican citizenship. When
the women arrived on the scene, Rousseau tells us, "*the dance was sus-
pended.*" And although the men "wanted to pick up the dance again, . . . *it was
impossible;* they did not know what they were doing any more; all heads were
spinning with a drunkenness sweeter than that of wine."[61] As Zerilli explains
it, "the presence of the women who came down to join the men (each woman
joins her husband) guards against another threat: the manly dance that might
very well have transgressed itself into homoerotic ecstasy."[62] The masculine
practice of the soldiers' dance must not accommodate the participation of
women, but their absent presence was necessary to keep men's focus off each
other, so that it can remain on the fatherland—a point we will explore more
completely in a moment.

What would happen if women began to engage in the practices constitutive
of masculine citizenship? Since gender is performatively constructed rather
than rooted in nature, it can never be permanently secured. In Zerilli's words,

what Rousseau teaches and fears is that natural man and woman are pedagogical
constructions and highly unstable ones at that. There is a profound sense in his
writings that gender boundaries must be carefully fabricated and maintained be-

cause they have no solid foundation in nature. . . . For what haunts the writer Rousseau above all else is the . . . fear that, if the code of gender difference is not strictly adhered to at each and every moment, all is lost.[63]

So if "women" began to engage in the practices constitutive of masculinity, then two things might happen. First of all, "women" might become "masculine"—which is scary enough. But even more frightening, "men" might become "feminine." Zerilli emphasizes this latter fear: What contemporary man fears is "the similitude of his sexual other"; he dreads "becoming woman." In other words, he dreads the femininity lurking inside himself. And as we recall, Hannah Pitkin came to a similar conclusion in her study of masculinity in the work of Machiavelli. As Pitkin puts it, "feminine power seems to be in some sense inside . . . men themselves."[64]

Traditionally, martial practices have functioned to construct a masculinity defined in direct opposition to femininity and to keep the feminine threat inside men at bay. As we recall, Pitkin emphasizes this in her discussion of Machiavelli's stress on military service in the militia: "only ferocious discipline and terrifying punishments" can prevent men from becoming feminine.[65] Likewise, Zerilli argues that in order to remain a masculine republican citizen, Rousseau's individual must not forgo "active participation in the public duties and ceremonies that alone safeguard against the feminine threat: military service . . . and the 'periodic assemblies.' "[66] Masculinity requires constant engagement in civic and martial practices.

And the same practices constitutive of masculinity also constitute republican citizenship. The republic depends upon constant civic and martial participation. In Rousseau's words,

as soon as public *service* ceases to be the main business of the citizens, and they prefer to serve with their pocketbooks rather than with their persons, the State is already close to ruin. Is it necessary to *march to battle?* They pay troops and stay home. Is it necessary to *attend council?* They name deputies and stay home. By dint of laziness and money, they finally have soldiers to enslave the country and representatives to sell it.[67]

For Rousseau, both republican citizenship and *armed masculinity* require continual participation in the civic and martial practices embodied by the ideal of the Citizen-Soldier.

Producing Republican Desire

One of the most important insights of the civic republican tradition—and one we should remember today—is that self-government aimed at the common good requires a passionate attachment to the community created through participation in civic practices. As I have argued, the figure of the Citizen-Soldier stands at the center of Rousseau's entire theoretical vision, because the civic and martial practices through which he is constituted create the *general will* and instill in individuals the characteristics that undergird their willingness to obey laws they give to themselves. In other words, while the *general will* yields legitimate laws that men respect on the basis of reason, this is not enough to ensure that men will obey the law. Men must also be moved to obey the law through a passionate connection to the republic and their fellow citizens, in other words, through patriotism, fraternity, and a common civic identity, all of which are created through engagement in civic and martial practices. Consequently, we can see in Rousseau's work that a *citizenship of civic practices* engages individuals both at the level of *reason* as they make laws for themselves and at the level of *passion* as they participate in the civic militia and its attendant communal festivities. In short, passion plays just as important a role in Rousseau's republican theory as does reason.

Feminist theorists often overlook the centrality of passion and the body to Rousseau's project—and to the Citizen-Soldier tradition of civic republicanism. For example, both Landes and Young accuse Rousseau of constructing a public sphere based on reason that requires the exclusion of passion. In Young's words, Rousseau (among others) "instituted a moral division of labor between reason and sentiment, identifying masculinity with reason and femininity with sentiment and desire." That is, "[b]y assuming that reason stands opposed to desire, affectivity and the body, the civic public must exclude bodily and affective aspects of human existence."[68]

This dichotomy profoundly affects citizenship, she continues, because it results in the exclusion "from the public [of] those individuals and groups that do not fit the model of the rational citizen who can transcend body and sentiment."[69] Landes makes a similar argument in her analysis of the Rousseauian "public sphere."[70]

What both Young and Landes fail to recognize, however, is that within Rousseau's theory—and we can see this in the civic republican tradition in general—citizenship engages passion and the body as well as reason and the mind. Citizens do not "transcend" affectivity and desire, but rather direct their emotions and passions toward the republic. That is, citizenship requires not only rational self-rule but also a passionate commitment to the republic, instilled in citizens as they engage together in civic and martial practices. Again, within Rousseau's work in particular and civic republicanism in general, citizenship engages individuals both at the level of reason and at the level of passion.[71]

However, while Rousseau recognized the centrality of passion to republican citizenship, he also realized the dangerous potential of passion to disrupt the fragile unity of the republic. That is, although passion could unify the republic if converted into patriotism, fraternity, and civic virtue, it could also disrupt the republic by fracturing the *general will* and drawing the citizen into the private realm, where he can focus on fulfilling his own personal desires. Rousseau's proposed solution to this puzzle entails the channeling of all desire toward the republic. For example, in the *Government of Poland,* Rousseau argues as follows:

> The newly born infant, upon first opening his eyes, must gaze upon the fatherland, and until his dying day should behold *nothing else.* Your true republican is a man who imbibed love of the fatherland, which is to say love of the laws and of liberty, with his mother's milk. That love makes up his *entire* existence; he has eyes *only* for the fatherland, lives *only* for his fatherland; the moment he is alone, he is a mere cipher; the moment he has no fatherland, he is no more; if not dead, he is worse off than if he were dead.[72]

From the moment he is born until the day he dies, the passion of the citizen must be directed toward the fatherland.

Rousseau praises Lycurgus for accomplishing this feat with the Spartans:

> He saw to it that the image of the fatherland was constantly before their eyes—in their laws, in their games, in their homes, *in their mating,* in their feasts. He saw to it that they never had an instant of free time that they could call their own. And out of this ceaseless constraint, made noble by the purpose it served, was born that *burning love of country* which was always the strongest—or rather *the only*—passion of the Spartans, and which transformed them into beings more than merely human.[73]

The fatherland should be the only passion of the republican citizen.

One of the keys to producing the fatherland as the central object of desire involves constant engagement in the civic and martial practices constitutive of masculine republican citizenship. So while the *general will* is not totalitarian, in that it reserves plenty of space for the individual outside the purview of the "state," Rousseau wants the individual to choose to spend all his free time participating in civic practices. In other words, the civic practices constitutive of the *general will* and republican citizenship are so time-consuming and all-encompassing that they leave little room to explore the particularistic, private interests excluded from the *general will.* In Rousseau's vision almost all of the individual's "particular desires, needs, and interests" should be aimed at the fatherland: "The better constituted the State, the more public affairs dominate private ones in the minds of the citizens."[74] The "life and freedom" the citizen maintains "independent" of "the body politic" should be absorbed with the civic practices constitutive of love of the fatherland.

Spending all his time participating in civic and martial practices, the individual develops a very strong civic identity that replaces all others. The citizen does not maintain or develop any more particular identities that could compete with his civic identity. Instead, each citizen "*gives his entire self.*"[75] He must not be torn by dual allegiances, because "everything that destroys social unity

is worthless."[76] Making the fatherland the primary object of desire and citizenship the only form of identification ensures republican unity.

The elimination of all competing identities and the elevation of the fatherland to the ultimate object of desire creates a very strong fusion among
Rousseau's citizens. Since the citizen's

> *alienation is made without reservation, the union is as perfect as it can be,* and no
> associate has anything further to claim. . . . *Each of us puts his person and all his*
> *power in common under the supreme direction of the general will; and in a body*
> *we receive each member as an indivisible part of the whole.*[77]

The comprehensive set of civic practices Rousseau advocates creates not only
a *general will* but also an extremely tight fusion of individuals into an "indivisible" union.

Despite Rousseau's democratic motivations, such a high degree of fusion is
ultimately undemocratic, because it requires the annihilation of difference and
the neglect of particular concerns. As Benjamin R. Barber puts it, "the more effective such affective institutions are, the less need there will be for democratic
politics, and the more likely it is that a community will take on the suffocating
unitary characteristics of totalistic states."[78] In fact, Rousseau exhibits three central characteristics of what Barber terms "unitary democracy." First, Rousseau
conceives of citizens as "brothers" rather than "neighbors."[79] Civic bonds are so
strong for Rousseau that civic festivities are like "the gathering of a big family."[80] Second, the identity of citizen is the only permissible one. And third, the
"ideal ground" of the community is "common beliefs, values, ends, identity
(substantive consensus)"—the unity of the *general will.* But while Rousseau incorporates these characteristics into the hidden foundation of his *general will,*
he does not embrace another key characteristic of "unitary democracy": the *citizenship of blood.*[81] We will discuss this point more fully later.

The Fantasy of Woman in Rousseau's Civic Imaginary

Rousseau's totalizing civic identity and "perfect" fusion depend upon the exclusion of women, because his plan to channel all desire toward the fatherland
depends upon the fantasmatic construction of Woman in his work—what Zerilli calls "that celestial object." Maintaining the fantasy of Woman requires
women's exclusion from the civic and martial practices constitutive of masculine republican citizenship, because if "women" participated in these practices,
the fragile performative construction of gender identity would be undermined
and with it the fantasy of Woman.

According to Zerilli, Rousseau constructs a fantasmatic Woman who demands virtue in men. This fantasy performs three important political functions.

In the first place, it transforms men's sexual desire for women into a desire for virtue. That is, if women are sexually attracted only to men of virtue, men must become virtuous republican citizens, if they hope to gain the love of their sexual objects. As Lord Bomston says to Saint Preux in *Julie:* "Do you know what has always made you love virtue? In your eyes it has taken on the form of that lovely woman who typifies it so well, and so dear an image could hardly let you lose the inclination for it."[82] Carol Blum and Elizabeth Wingrove concur with Zerilli: Rousseau eroticizes virtue as a way of enticing men to act like men.[83] Men enjoy the power actresses exercise over them in the theater, Rousseau reasons, so why not use men's desire to be under the spell of women for the good of the republic? Rousseau wants republican festivals to replace the actress as object of men's voyeuristic desires. "So 'magnificent' is this [republican] spectacle of men," Zerilli tells us, "that it will extinguish man's fatal desire to gaze at that other blazing magnificence: the sumptuous body of the salonniere or the actress."[84] Or as Wingrove puts it, "Rousseau's strategy is to nourish desires whose satisfaction consists in the fulfillment of republican duty."[85] In this way, Rousseau tries to put sexual passion into the service of the fatherland.

Secondly, Zerilli argues that Rousseau constructs a fantasy of the virtuous Woman to help protect men from their desire for actual women. That way, when a woman does not live up to the fantasy in a man's head, he will no longer desire her. In Zerilli's words, "that celestial object, that magnificent fetish, the imperious and mute woman of the male imaginary . . . *protects man against that other sort of woman and all her sex,* against the speaking woman of the theater and the salon."[86] As Rousseau himself says, "by providing the imaginary object, . . . I easily prevent my young man from having illusions about real objects."[87] That is,

> it is unimportant whether the object [the fantasy of Woman] I depict for him [Emile] is imaginary; it suffices that it make him disgusted with those that could tempt him; it suffices that he everywhere find comparisons which make him prefer his chimera to the real objects that strike his eye. And what is true love itself if it is not chimera, lie, and illusion? We love the image we make for ourselves far more than we love the object to which we apply it.[88]

In love with the fantasy of Woman, men can reject any woman who falls short of demanding republican virtue in him. The fantasy of Woman becomes the device Rousseau uses in his scheme to channel passion into the service of the republic.

And finally, Zerilli argues that the fantasy of Woman who demands manly republican virtue in men protects man against his own feminine desires, which would soften his martial virtue and hinder his ability to become a citizen-soldier.[89] As Rousseau tells an Emile on the brink of manhood: "A new enemy is arising which you have not learned to *conquer* and from which I can no longer save you. This enemy is yourself."[90] The enemy is inside Emile: It is not only "the desire for a woman" but also the desire "to be at the feet of woman, if not to be a woman."[91] The fantasy of the virtuous Woman demands that men act like masculine republican citizen-soldiers and so helps prevent men from submitting to feminine desires.

In other words, not only does Rousseau's fantasy of Woman help eroticize the civic practices constitutive of citizen-soldiering, but the presence of the fantasy as an object of men's desires helps prevent the civic passion expressed in male-only groups from slipping into explicit homoeroticism. If men started focusing on homoerotic desire, two things would happen, both of which would lead to the collapse of the republic. First of all, explicit homoerotic desire could disrupt the unity of the republic, by creating divisions among men. Secondly, if men gave in to "feminine" desires, the fragile performative construction of gender identity would begin to break down, and without "masculinity" there could be no "republican citizenship." If masculinity is constructed in opposition to femininity, then to give over to feminine desires in oneself would be to abandon masculinity, which would be to abandon republican citizenship, since the two are linked in Rousseau's thought via the ideal of the Citizen-Soldier.

But while Rousseau's recognition of the precariousness of performatively constructed categories of civic and gender identity led him into a "quick retreat into a rigid conception of sexual difference,"[92] his *citizenship of civic practices* actually leaves open the radical possibility of expanding republican citizenship to include "women." As we have seen so far, engagement in martial practices constitutes *armed masculinity,* which is then fused onto republican citizenship through the ideal of the Citizen-Soldier. *Armed masculinity*—or any type of masculinity for that matter—is not a *natural* attribute of male individuals. Instead, it is an always-precarious artifice that must be constantly constructed and reconstructed through engagement in masculine practices. Civic republican citizenship has been masculinized by its connection to soldiering, which has been masculinized by its cultural conflation with masculinity.

But if masculinity is not a natural attribute of male individuals, but instead is constructed through a series of performances, then what would happen if "women" began to engage in the practices constitutive of masculinity? Throughout his work, Rousseau vehemently insists that women be denied access to participation in the practices constitutive of republican citizenship and *armed masculinity*—such as the soldiers' dance discussed earlier. So while masculine republican citizens have been constituted through engagement in civic and martial practices, feminine subjects have been (partially) constituted through the exclusion from these same practices. If this is the case, then what would happen if "women" began to engage in the practices constitutive of both *armed masculinity* and republican citizenship? Could women's transgressive engagement in these practices render them citizen-soldiers?

The *performativity theory* of identity inherent in Rousseau's vision leaves open the possibility of *subversive transgender performances.* In other words, if gender does not flow naturally from biological sex, then there is no guarantee that a biological male will become *masculine* and a biological female *feminine*. For example, Blum argues that, in his fantasies, Rousseau often identified as a woman: "He began imagining himself to be alternately Julie, Claire, and Saint Preux. . . . In his ecstasy Rousseau could move easily back and forth between subject and object, male and female, lover and beloved."[93] Moreover, the culturally constructed imperative that "men" must become "masculine" and "women" become "feminine" does not always succeed—a fact of which Rousseau was well aware and on which he commented often. What I am suggesting here is that the possibility of transgender identification highlights the artificiality of supposedly "natural" categories of gender.

Politically, this means that "women's" participation in the civic and martial practices culturally deemed "masculine" could radically undermine the traditional dichotomous construction of gender and the sexism it generates. In other words, "women's" transgressive performance of the behaviors constitutive of citizen-soldiers should work to undermine the idea that "men" and "women" must be restricted by the cultural imperatives of "masculinity" and "femininity," respectively. Moving beyond restrictive gender norms will allow all individuals the freedom to live as they choose and the opportunity to become republican citizens. The movement beyond the traditional dichotomous constructions of gender not only undermines sexism but could also allow full civic subjectivity for "women." That is, it could open up the possibility for "women" to become republican citizens on an equal basis with "men."

Nevertheless, despite the possibility of *subversive transgender performances* inherent in the *citizenship of civic practices,* Rousseau's particular rendition of the civic republican tradition truncates the radical possibility of extending republican citizenship to "women." That is to say, in constructing his version of civic unity, Rousseau uses the fantasy of Woman as a way of channeling all passion toward the fatherland. And as I have argued, this fantasy requires the exclusion of biological females from the civic and martial practices. "Women" must be excluded because their exclusion plays a key role in the performative construction of gender difference that underlies the possibility of the fantasmatic Woman. Without gender difference, there can be no fantasy of Woman to help men channel their erotic desires toward the fatherland. And, more mundanely, men could not continue to maintain unrealistic fantasies about women if they were interacting with them as equals in both political and military spheres. "Women" cannot be included in Rousseau's version of republican citizenship, not because his civic and martial practices are *inherently* "masculine," but rather because he uses the fantasy of Woman in his (futile) attempt to channel sexual desire toward the fatherland, and this attempt to create a "perfect" fusion among his citizens requires the continuation of the traditional configuration of gender.

Rousseau does not rely primarily on the real or imagined threat of an external enemy to unify his citizenry.[94] Instead he unifies his citizenry to some extent by having each man struggle against two internal enemies present within everyone:

particularity and femininity. Both Hannah Arendt[95] and Linda Zerilli[96] discuss the role of internal enemies in Rousseau's system. Rousseau's citizen-soldiers must purge all particularity and femininity from themselves in order to maintain civic unity. At the same time, however, while Rousseau does make some use of the idea of "internal enemies," for the most part he creates civic unity by envisioning an all-encompassing set of civic and martial practices that produce not only a totalizing civic identity that leaves no room for particularity but also versions of the *general will,* fraternity, and patriotism so strong that they easily slip over into homogeneity, fusion, and, as we shall see momentarily, nationalism.

Rousseau's Citizen-Soldier and the Problem of Nationalism

Finally, Rousseau's vision of republican unity, constructed through a time-filling set of civic and martial practices, includes a version of patriotism so strong that it easily slips toward nationalism. Rousseau rightly recognizes that a strong passionate attachment to the fatherland and its people cannot be endlessly inclusive. Opposing cosmopolitanism, Rousseau argues for the virtue of the small republic. A vast empire cannot engender the feelings of patriotism and fraternity necessary to republican self-government, because a person cannot be attached to the whole world: "The people has less affection . . . for the homeland which is like the whole world in its eyes, and for its fellow citizens, most of whom are foreigners to it."[97] On this point Rousseau again stands outside of the rationalist paradigm. A politics based on rational discourse can be infinitely expanded—that's one of its primary virtues. Civic republican politics cannot be, however, because of the role it gives to passion in the creation of citizenship.

So while Rousseau's overemphasis on *commonality among citizens* produces fusion, a totalizing civic identity, and homogeneity out of fraternity, civic virtue, and the *general will,* his overemphasis on the *particularity of different peoples* produces nationalism out of patriotism. Rousseau begins by arguing that the existence of a distinctive civic identity underlies the possibility of masculine citizen-soldiering:

> There is one rampart, however, that will always be readied for its defense, and that no army can possibly breach; and that is the virtue of its citizens, their patriotic zeal, *in the distinctive cast that national institutions are capable of impressing upon their souls.* See to it that every Pole is incapable of becoming a Russian, and I answer for it that Russia will never subjugate Poland.[98]

Bemoaning the fact that "there is no such thing nowadays as Frenchmen, Germans, Spaniards, or even Englishmen—only Europeans," Rousseau argues that "*national* institutions" are needed to give "form to the genius, the character, the tastes, and the customs of a people."[99] According to Rousseau, the existence of the republic, its laws, and its liberty depends upon a particular people's developing a distinctive civic identity.

Rousseau's advocacy of a distinctive civic identity, however, can easily yield the chauvinism and xenophobia characteristic of nationalism. For example, Rousseau wanted the Poles to be "purely Polish." He wanted to "endear Poland to its citizens and develop in them an instinctive distaste for mingling with the peoples of other countries." A particular set of Polish civic practices must make "life in Poland . . . more fun than life in any other country, but not the same kind of fun."[100] That is, one must

> give a different bent to the passions of the Poles; in doing so, you will shape their minds and hearts in a national pattern that will set them apart from other peoples, that will keep them from being absorbed by other peoples, or finding contentment among them, or allying themselves with them.[101]

Rousseau praises Moses, who by determining

> that his people should never be absorbed by other peoples, . . . devised for them customs and practices that could not be blended into those of other nations and weighted them down with rites and peculiar ceremonies. . . . Each fraternal bond that he established among the individual members of his republic became a further barrier, separating them from their neighbors and keeping them from becoming one with those neighbors.[102]

The strong sense of particularity and the distinctive and all-encompassing civic identity that Rousseau advocates unfortunately laid the groundwork for the emergence of nationalism and its imagined spiritual unity.

Nevertheless, while Rousseau's emphasis on "*national* institutions" as essential to creating particular peoples with particular genius, character, tastes, and customs, who are themselves "rather than other people," and who have an "ardent love of fatherland"[103] certainly could lead to chauvinism, xenophobia, and nationalism,[104] his emphasis on civic institutions and practices as absolutely necessary to the production of citizens indicates his advocacy of a *citizenship of civic practices* rather of than a *citizenship of blood* (*ius sanguinis*). That is, Rousseau does not argue that Poles should be citizens because they have Polish blood. Instead he argues that without engaging in civic practices, they will not be citizens at all. Thus, he also rejects the standard liberal notion of citizenship (*ius solis*): Poles should not be citizens by virtue of their residence in Poland but, again, only through engagement in civic practices. Moreover, while Rousseau lays the groundwork for nineteenth-century nationalistic developments, he does stop short of actually positing the spiritual unity of different peoples—an essential characteristic of nationalism and its progeny, fascism.

Rousseau recognizes that the virtue of patriotism and the vice of nationalism are inextricably linked. As he says, without distinctive national institutions, "national hatreds will die out, but so will love of country."[105] The problem for democratic theorists is always how to maintain the virtue while minimizing the vice. Rousseau avoids the slide toward conquest that comes out of Machiavelli's version of the Citizen-Soldier tradition. However, despite his best efforts, Rousseau ends up bolstering the other side of nationalism's foundation—the distinctiveness of different peoples.

As we saw with Machiavelli, despite the theorist's best efforts at constructing a dialectical theory of civic republican citizenship, in the realm of realpolitik, the theoretical dialectic often breaks down. Rousseau's synthesis of the individual and the community at first inspired the democratic French Revolution, but the fragile theoretical construction broke down under the pressure of war. An excellent example of the disintegrative effects of realpolitik on grand theoretical syntheses is the transformation of the famous *levée en masse* during the course of the French Revolution. The *levée en masse* began as a republican ideal that emerged directly out of the Citizen-Soldier tradition. As originally formulated, the *levée en masse* linked military service to participatory citizenship. R. R. Palmer explains that although "the term *levée en masse* has become frozen to signify the universal military service of the Revolution, a conscription conducted by government and designed to expel foreign invaders . . . in its origin the term meant much more."[106] Translated literally as "mass rising," the term originally connoted more than just universal military service; it also included the

> *general rising of the people for any purpose,* with or without the assistance of official persons who did not command much public confidence. It could be *a swarming of citizen soldiers to defy the regular armies of Prussia and Austria.* It could be *a rising of the sections of Paris* against the Convention or some of its members. It could be *an armed insurrection or an unarmed demonstration in the streets.*[107]

In 1793, the *levée en masse* meant civic actions of all kinds, including both participation in assemblies and the bearing of arms.

The *levée en masse* linked military service with participatory citizenship. During the Revolution, political enfranchisement and military conscription advanced hand in hand. In order to recruit men to fight in the war, the government extended democracy to the popular masses.[108] However, during the course of the Revolution, military service and participatory citizenship became uncoupled. Although universal conscription of men continued throughout the war, resulting by 1794 in an army of almost a million, the other essential half of this Citizen-Soldier pairing, participatory citizenship, was increasingly undermined. The Citizen-Soldier ideal was reduced to its less democratic half: universal conscription.

Popular Sovereignty and *Subversive Transgender Performances*

But whereas Rousseau's theory of the Citizen-Soldier fed directly into the nationalism ushered in by the French Revolution—complete with its own set of virtues and vices—Rousseau's theory of the *general will* forms the foundation for the radical democratic idea of popular sovereignty, an ideal that animated the democratic strands of both the French and the American revolutions.[109] The radical republicans of the French Revolution—the *sans-culottes*—clearly understood citizenship as requiring engagement in civic and martial practices as embodied by the ideal of the Citizen-Soldier. Citizenship meant both to act and to fight.[110] The *sans-culottes,* women as well as men, had a very concrete idea of both popular sovereignty and republican citizenship: They both meant "deliberating in . . . section assemblies, bearing arms, [and] sitting on the Revolutionary Tribunal." The radical republicans understood citizenship as entailing the direct participation of all citizens in political assemblies, the accountability of elected representatives to the citizenry, and the ability of citizens to enforce the law and to revolt if the law was violated by its delegates. Thus the "second" Revolution of 1792 clearly represents the Citizen-Soldier tradition at its most radical: Radical republican men *and* women embraced a notion of republican citizenship that required both participation in legislation for the common good and service in the civic militia.[111]

The armed processions during the spring of 1792 that paved the way for the *levée en masse* mark the beginning of the most radical phase of women's struggle to inhabit the category of the Citizen-Soldier. Recent scholarship shows that women unofficially engaged in the practices of masculine republican citizen-soldiering, including both participation in political assemblies and the bearing of arms. Radical republican women actually bore arms throughout the Revolution, lobbied the National Assembly for permission to form a Women's National Guard, donned the attire that signified male republican citizenship, and unofficially voted on the Constitution of the Year I, sending word to the male leadership that they approved. For these women the right to bear arms to defend the republic was absolutely central to their struggle for equal citizenship.[112]

Even before the radical republican "second" Revolution began in full, "women of the people" understood the centrality of arms to republican citizenship. As Pauline Leon barricaded the streets of Paris alongside men on July 14, 1789, she armed herself with the pike just as they did. And of course there is the famous "Women's March on Versailles" that marked "a transitional moment in the transformation of subjects into a militant citizenry identifying itself as the sovereign nation." During this march women's engagement in civic practices, including the bearing of arms, helped bolster "remarkably broad demands for the political and military status and rights of female citizenship."[113] Women clearly hoped to attain the status of republican citizenship through their engagement in the "masculine" practices of citizen-soldiers.

The "women of the people" clearly recognized that if they were to be republican citizens, they must be allowed to bear arms alongside men. On March 6, 1791, Pauline Leon presented a petition with over 300 signatures to the National Assembly claiming the right to bear arms in defense of the republic. A lengthy excerpt from her speech follows:

Patriotic women come before you to claim the right which any individual has to *defend his life and liberty*. . . . Yes, Gentlemen, *we need arms,* and we come to ask your permission to procure them. May our weakness be no obstacle; *courage* and intrepidity will supplant it, and *the love of the fatherland and hatred of tyrants* will allow us to brave all dangers with ease. Do not believe, however, that our plan is to abandon the care of our families and home, always dear to our hearts, to run to meet the enemy. No, Gentlemen. We wish only to defend ourselves the same as you; you cannot refuse us, and society cannot deny the right nature gives us, unless you pretend the Declaration of Rights does not apply to women, and that they should let their throats be cut like lambs, without the right to defend themselves. For can you believe the tyrants would spare us? No, no—they remember October 5 and 6, 1789. . . . Why then not terrorize aristocracy and tyranny with all the resources of civic effort [*civisme*] and the purest zeal, zeal which is only the natural result of *a heart burning with love for the public weal*? . . . The rages and plots of aristocrats . . . will not succeed in vanquishing a whole people of united brothers *armed to defend their rights*. We also demand only the honor of sharing their exhaustion and glorious labors and of *making tyrants see that women also have blood to shed for the service of the fatherland in danger. . . . We hope to obtain . . . permission to procure pikes, pistols, and sabres (even muskets for those who are strong enough to use them), within police regulations . . . [and] permission to assemble on festival days and Sundays on the Champ de la Federation, or in other suitable places, to practice maneuvers with these arms.*[114]

Clearly, participation in martial as well as civic practices was viewed by these women as central to achieving republican citizenship.

Despite Leon's reference to motherhood, her struggle for citizenship should be understood as motivated not *simply* by the desire to fulfill her traditional familial duties more effectively but also by the desire to act as a citizen in the participatory republican sense. This interpretation differs from much of the traditional scholarship that views women's activism during the French Revolution as revolving around typical feminine domestic issues, such as subsistence. Such arguments reduce women's political activism to being merely an extension of their responsibility for the domestic sphere. Although domestic concerns no doubt played some role, more recent historical analysis stresses the *political* nature of women's activism. For example, Dominique Godineau argues that the women who engaged in the practices of militant republican citizenship were motivated by political demands as well as subsistence concerns: Women of the people wanted 'Bread, but not at the price of liberty,' 'Bread and the Constitution of 1793.' As Godineau explains:

If women of the people did not occupy center stage when the subsistence question was secondary, that does not in any sense signal their absence. They simply merged into the larger whole of the popular masses; they were present not as collectivities of women, but rather as individuals of the feminine sex.[115]

Women engaged in political action alongside men and had the same understanding of republican citizenship as they did: participatory citizenship and service in the civic militia. Since both sets of practices were necessary to the constitution of the republican Citizen-Soldier, women knew that they too must engage in martial as well as civic practices, if they were to become republican citizens.

The struggle of women to gain republican citizenship on an equal basis with men culminated in the founding in 1793 of *La Societe des Citoyennes Republicaines Revolutionaires* by Pauline Leon and Claire Lacombe. Its regulations stated that the *Societe* was a "family of sisters" in which "no member can be denied the right to speak" in its deliberative proceedings and that its "purpose is to be armed to rush to the defense of the Fatherland."[116] Not only did these radical women claim the right to engage in traditionally masculine civic and martial practices, but they also assumed male attire—pantaloons and the red liberty cap—and strode armed through the streets of Paris, enforcing the revolutionary mandate that all women wear the tricolor cockade and the male liberty cap—emblems of republican citizenship.

This transgressive engagement in masculine practices was seen as threatening to the always-fragile constructions of femininity and masculinity. Many revolutionary republicans—female as well as male—found women's assumption of male attire particularly disturbing. For example, one citoyenne was recorded as saying that she told two women wearing les bonnets rouges, "Off with *les bonnets rouges* [red caps], because they are only for men to wear."[117] According to another document, "all [the market] women were in agreement that violence and threats would not make them dress in a costume [which] they respected but which they believed was intended for men."[118] In the words of Commune president Chaumette,

> it is contrary to all the laws of nature for a woman to want *to make herself a man.* The Council must recall that some time ago these *denatured* women, these *viragos,* wandered through the markets with the red cap to sully that badge of liberty and wanted to force all women to take off the modest headdress that is appropriate for them. The place where people's magistrates deliberate must be forbidden to every person who insults nature.[119]

Implicit in this statement is some understanding of the importance of "performativity" in the constitution of gender identity.

The specter of women assuming men's attire seemed to be particularly threatening to many of the revolutionaries—women as well as men. In fact, not long after this radical assumption of masculine attire, all women's associations were banned. In purging women from the revolutionary republic, very strong appeals were made to gender difference and women's natural place as virtuous republican mothers.[120] This is an interesting phenomenon in light of our discussion of Rousseau and some recent theoretical work on cross-dressing. For instance, Judith Butler has suggested that "drag" demonstrates "the mundane way in which genders are appropriated, theatricalized, worn, and done; it implies that all gendering is a kind of impersonation and approximation."[121] Thus, cross-dressing highlights the precariousness of the ostensibly natural constructions of gender. Similarly, Marjorie Garber argues that the appearance of the "transvestite" in culture signals a "category crisis," which she defines as "a failure of definitional distinction, a borderline that becomes permeable, that permits of border crossings from one (apparently distinct) category to another: black/white, Jew/Christian, noble/bourgeois, master/servant, master/slave."[122] Cross-dressing thus challenges "easy notions of binarity, putting into question the categories of 'female' and 'male,' whether they are considered essential or constructed, biological or cultural."[123]

The image of radical republican women in "drag" called into question the naturalness of gender difference and hence the justification for excluding "women" from republican citizenship. Thus, the case of political struggle of *la Societe des Citoyennes Republicaines Revolutionaires* for republican citizenship during the French Revolution constitutes a historical moment through which we can begin to imagine the possibility of *subversive transgender performances* opening up the category of republican citizenship to all people as well as transforming the traditional configurations of gender that historically underwrote "women's" exclusion from citizenship. This possibility offers hope to feminists who want to conceptualize a citizenship that can be truly inclusive and yet do not want to give up on the discourse of civic republicanism because it bolsters calls for a strong notion of participatory citizenship, popular sovereignty, and government for the good of all.

In conclusion, Rousseau's version of the Citizen-Soldier tradition presents a package of interconnected virtues and vices. But although his work took a conceptual giant step toward nationalism—complete with its own set of virtues and vices—he also theorized the radical ideal of popular sovereignty that animated the most democratic aspects of the French Revolution, including women's transgressive struggle to occupy the category of "republican citizen-soldier." Though those women's efforts ultimately failed, the potential for *subversive transgender performances* still remains inherent in a *citizenship of civic practices* that roots identity in practices rather than in nature.

Notes

1. Jean-Jacques Rousseau, *The Government of Poland,* trans. Willmoore Kendall (Indianapolis: Hackett, 1985), 81.
2. Rousseau, *Government of Poland,* 80–81.
3. Rousseau, *Government of Poland,* 81.
4. Rousseau, *Government of Poland,* 80.

5. Rousseau, *Government of Poland,* 85.

6. Jean-Jacques Rousseau, *On the Social Contract,* ed. Roger D. Masters, trans. Judith R. Masters (New York: St. Martin's Press, 1978), 54, emphasis mine.

7. Rousseau, *Social Contract,* 62.

8. "Instantly, in place of the private person of each contracting party, this act of association produces a moral and collective body, composed of as many members as there are voices in the assembly, which receives from this same act its unity, its common *self,* its life, and its will." Rousseau, *Social Contract,* 53–54, emphasis mine.

9. Rousseau, *Social Contract,* 59.

10. On this point I disagree with Iris Marion Young, who criticizes Rousseau's citizen for being an "impartial moral reasoner . . . who stands outside of and above the situation about which he or she reasons, with no stake in it, or is supposed to adopt an attitude toward a situation as though he or she were outside and above it" (60). With this accusation, Young not only confuses Rousseau with Kant, but also misses Rousseau's *citizenship of civic practices.* That is to say, Rousseau's citizen cannot stand outside or above politics because he is only constituted through engagement political practices. See Iris Marion Young, "Impartiality and the Civic Public," in *Feminism as Critique,* ed. Seyla Benhabib and Drucilla Cornell (Minneapolis: University of Minnesota Press, 1988).

11. Young, "Impartiality and the Civic Public," 61.

12. Michael Walzer makes this point in *What It Means to Be an American: Essays on the American Experience* (New York: Marsilio Publishers, 1996), 13.

13. Rousseau, *Social Contract,* 53–54, emphasis mine.

14. The entire quote reads: "There is a third, more bizarre, type of religion [Roman Catholicism] which, by giving men two legislative systems, two leaders, and two homelands, subjects them to contradictory duties, and prevents them from being simultaneously devout men and citizens. . . . The third is so manifestly bad that it is a waste of time to amuse oneself by proving it. Everything that destroys social unity is worthless. All institutions that put man in contradiction with himself are worthless." Rousseau, *Social Contract,* 128.

15. Rousseau, *Government of Poland,* 61.

16. Rousseau, *Government of Poland,* 58.

17. Rousseau, *Government of Poland,* 58, emphasis mine.

18. Rousseau, *Government of Poland,* 76.

19. See Young, "Impartiality and the Civic Public," and Joan Landes, *Women and the Public Sphere in the Age of the French Revolution* (Ithaca: Cornell University Press, 1988).

20. Rousseau, *Social Contract,* 108.

21. Young, "Impartiality and the Civic Public," 62.

22. J. L. Talmon, *The Origins of Totalitarian Democracy* (London: Secker & Warburg, 1952).

23. Talmon, *The Origins of Totalitarian Democracy,* 53.

24. Talmon, *The Origins of Totalitarian Democracy,* 56.

25. Rousseau, *Government of Poland,* 4.

26. Rousseau, *Social Contract,* 77.

27. Rousseau, *Government of Poland,* 5, emphasis mine.

28. Rousseau, *Government of Poland,* 4.

29. Rousseau, *Government of Poland,* 22.
30. Jean-Jacques Rousseau, *Politics and the Arts: Letter to M. d'Alembert on the Theatre,* trans. Allan Bloom (Ithaca: Cornell University Press, 1960), 130.
31. Rousseau, *Letter to M. d'Alembert,* 126.
32. Rousseau, *Government of Poland,* 131, emphasis mine.
33. Mona Ozouf, *Festivals and the French Revolution,* trans. Alan Sheridan (Cambridge, MA: Harvard University Press, 1988), 9.
34. Rousseau, *Letter to M. d'Alembert,* 135, emphasis mine.
35. Rousseau, *Letter to M. d'Alembert,* 135, emphasis mine.
36. Rousseau, *Letter to M. d'Alembert,* 135, emphasis mine.
37. Jean-Jacques Rousseau, *Discourse on the Sciences and Arts (First Discourse),* ed. Roger D. Masters, trans. Roger D. and Judith R. Masters (New York: St. Martin's Press, 1964), 55.
38. Rousseau, *Letter to M. d'Alembert,* 102.
39. Rousseau, *First Discourse,* 37.
40. Rousseau, *First Discourse,* 54.
41. Rousseau, *First Discourse,* 55.
42. Rousseau, *First Discourse,* 45.
43. Rousseau, *First Discourse,* 56, 50, 45.
44. Rousseau, *First Discourse,* 45.
45. Rousseau, *First Discourse,* 59.
46. Rousseau, *Letter to M. d'Alembert,* 82.
47. Judith Shklar, *Men & Citizens: A Study of Rousseau's Social Theory* (Cambridge: Cambridge University Press, 1969). My analysis builds on Linda Zerilli's departure from Shklar's classic interpretation: "Contesting the critical consensus that Rousseau presents us with the choice of making either a man or a citizen (since one cannot make both at once), I show that to be the latter one must, in the first place, be the former, and that to be a man is to be no more a product of nature than is to be a citizen to be a denatured man." See Linda Zerilli, *Signifying Woman: Culture and Chaos in Rousseau, Burke, and Mill* (Ithaca: Cornell University Press, 1994), 18.
48. Shklar, *Men & Citizens,* 3.
49. "The wish to play a public role, to develop one's civic capacities, to belong to a purposeful order, to take part in an organized drama, is as much a part of a morally adult life as the desire to be a self-sufficient whole, united only with those whom one loves and independent of all that interferes with one's real needs. Choose, however, one must, or rather ought, even though one never does." Shklar, *Men & Citizens,* 32.
50. Zerilli, *Signifying Woman,* 28, emphasis mine.
51. Elizabeth Wingrove, "Sexual Performance as Political Performance," *Political Theory* 23, no. 4 (November 1995): 588, emphasis mine.
52. Rousseau, *Letter to M. d'Alembert,* 102–3, emphasis mine.
53. See Judith Butler, *Gender Trouble: Feminism and the Subversion of Identity* (New York: Routledge, 1990), 33.
54. Rousseau, *First Discourse,* 36.
55. Rousseau, *Letter to M. d'Alembert,* 103, emphasis mine.
56. Rousseau, *First Discourse,* 57, emphasis mine.
57. Rousseau, *First Discourse,* 45, 43.
58. Wingrove, "Sexual Performance," 609.

59. Rousseau, *Letter to M. d'Alembert,* 130.
60. Rousseau, *Letter to M. d'Alembert,* 112.
61. Rousseau, *Letter to M. d'Alembert,* 135–36.
62. Zerilli, *Signifying Woman,* 38.
63. Zerilli, *Signifying Woman,* 18.
64. Hannah Fenichel Pitkin, *Fortune Is a Woman: Gender and Politics in the Thought of Niccolo Machiavelli* (Berkeley: University of California Press, 1984), 136.
65. Pitkin, *Fortune Is a Woman,* 136.
66. Zerilli, *Signifying Woman,* 57.
67. Rousseau, *Social Contract,* 101–2, emphasis mine.
68. Young, "Impartiality and the Civic Public," 66.
69. Young, "Impartiality and the Civic Public," 66.
70. See Landes, *Women and the Public Sphere.*
71. One of the reasons both Landes and Young overlook this point is because they both wrongly conflate Rousseau's work with Jurgen Habermas's. But Rousseau and Habermas stand within very different traditions: Habermas stands within the German Idealist tradition, the origins of which lay with Kant. And while Habermas improves on Kant by switching from monological to dialogical reason, he does not accept Rousseau's belief that passion is just as critical to citizenship as is reason. In fact, I would argue that one of the key characteristics that separates civic republicanism from German Idealism (and liberalism too for that matter) is the emphasis on bringing passion as well as reason into the service of the republic.
72. Rousseau, *Government of Poland,* 19, emphasis mine.
73. Rousseau, *Government of Poland,* 7, emphasis mine.
74. Rousseau, *Social Contract,* 102.
75. Rousseau, *Social Contract,* 53.
76. Rousseau, *Social Contract,* 128.
77. Rousseau, *Social Contract,* 53.
78. Benjamin R. Barber, *Strong Democracy: Participatory Politics for a New Age* (Berkeley: University of California Press, 1984), 243.
79. Barber, *Strong Democracy,* 219.
80. Rousseau, *Letter to M. d'Alembert,* 131.
81. Barber, *Strong Democracy,* 219.
82. Jean-Jacques Rousseau, *La Nouvelle Heloise: Julie, or the New Eloise,* trans. and abridged Judith H. McDowell (University Park: Penn State Press, 1968), 343.
83. See Carol Blum, *Rousseau and the Republic of Virtue: The Language of Politics in the French Revolution* (Ithaca: Cornell University Press, 1986), chap. 5; Zerilli, *Signifying Woman,* chap. 2; and Wingrove, "Sexual Performance as Political Performance."
84. Zerilli, *Signifying Woman,* 37.
85. Wingrove, "Sexual Performance," 596–97.
86. Zerilli, *Signifying Woman,* 18.
87. Jean-Jacques Rousseau, *Emile; or, On Education,* ed. Allan Bloom (New York: Basic Books, 1979), 329.
88. Rousseau, *Emile,* 329.
89. Zerilli, *Signifying Woman,* 18.
90. Rousseau, *Emile,* 431.

91. Zerilli, *Signifying Woman*, 40.
92. Zerilli, *Signifying Woman*, 18.
93. Blum, *Rousseau and the Republic of Virtue*, 61–62.
94. Hannah Arendt argues to the contrary: "Politically speaking, [Rousseau] presupposed the existence and relied upon the unifying power of the common national enemy. Only in the presence of the enemy can such a thing as *la nation une et indivisible*, the ideal of French and of all other nationalism, come to pass." See *On Revolution* (London: Penguin Books, 1963), 77.
95. Arendt argues that Rousseau "wished to discover a unifying principle within the nation itself that would be valid for domestic politics as well. Thus, his problem was where to detect a common enemy outside the range of foreign affairs, and his solution was that such an enemy existed within the breast of each citizen, namely, in his particular will and interest; the point of the matter was that this hidden, particular enemy could rise to the rank of a common enemy—unifying the nation from within—if one only added up all particular wills and interests. The common enemy within the nation is the sum total of the particular interests. . . . In Rousseau's construction, the nation need not wait for an enemy to threaten its borders in order to rise 'like one man' and to bring about the *union sacree;* the oneness of the nation is guaranteed in so far as each citizen carries within himself the common enemy as well as the general interest which the common enemy brings into existence; for the common enemy is the particular interest or the particular will of each man." See *On Revolution*, 77–78.
96. Zerilli notes that Rousseau tells Emile: "Dear Emile, it is in vain that I have dipped your soul in the Styx; I was not able to make it everywhere invulnerable. *A new enemy is arising which you have not learned to conquer* and from which I can no longer save you. *This enemy is yourself*" (431). Building on this quote, Zerilli argues that Rousseau "prepare[s] the child for battle with 'the enemy' who will appear in Book V: the desire for a woman, to be at the feet of woman, if not to be a woman." See *Signifying Woman*, 40.
97. Rousseau, *Social Contract*, 72.
98. Rousseau, *Government of Poland*, 11, emphasis mine.
99. Rousseau, *Government of Poland*, 11.
100. Rousseau, *Government of Poland*, 14.
101. Rousseau, *Government of Poland*, 12.
102. Rousseau, *Government of Poland*, 6.
103. Rousseau, *Government of Poland*, 11.
104. A point emphasized by Shklar.
105. Rousseau, *First Discourse*, 38.
106. R. R. Palmer, *The Age of the Democratic Revolution*, vol. 2 (Princeton: Princeton University Press, 1964), 103.
107. Palmer, *Age of the Democratic Revolution*, vol. 2, 104.
108. Palmer, *Age of the Democratic Revolution*, vol. 2, 15.
109. See Palmer, *The Age of the Democratic Revolution*.
110. In the following passage, historian R. R. Palmer interweaves descriptions of participatory citizenship with descriptions of the bearing of arms. These two sets of practices are linked by the sans-culottes because they are linked within the larger Citizen-Soldier tradition of civic republicanism within which the "popular democrats" were

situated: "The sans-culottes were in effect popular democrats. In the crisis and break-down of 1792 they represented an enormous wave of citizen self-help. They applied the great concepts of liberty, equality, and the sovereignty of the people to themselves and the concrete circumstances with which they were personally familiar. They believed that they themselves were sovereign, in face-to-face contact in their section meetings; and that distant elected persons were only their delegates, often not to be trusted. They favored what a later generation in America would know as referendum and recall. 'Consent of the people' meant their consent in their own assemblies. The right to bear arms meant that they should carry pikes in their own streets. The judgment of the people meant that they should denounce their own neighbors for suspicious behavior or unsuitable sentiments, and that their own committees should put them under arrest. They resisted attempts at control of their activities by the Convention and its Committee of Public Safety in 1793.

"If they presumed to exercise sovereignty, they accepted the corresponding responsibilities; they were ready to give their time, to act and to fight. The younger ones were gradually absorbed into the army. They spent long hours at meetings, and in the work of committees, or on the exposure of suspects, or on errands and missions and patrols about the city, or in exchange of delegations with sister groups, or in semimilitary formations in which men from the city went into rural areas to procure food from the peasants, or bring patriotic pressure to bear in other communities." See Palmer, *Age of the Democratic Revolution,* vol. 2, 46–47.

111. See Palmer, *Age of the Democratic Revolution,* vol. 2, 46–47.

112. Darline Levy and Harriet B. Applewhite, "Women and Militant Citizenship in Revolutionary Paris," in *Rebel Daughters: Women and the French Revolution,* ed. Sara E. Mezler and Leslie W. Rabine (New York: Oxford University Press, 1992); and "Women, Radicalization and the Fall of the French Monarchy," in *Women and Politics in the Age of the Democratic Revolution,* ed. Harriet B. Applewhite and Darline Levy (Ann Arbor: University of Michigan Press, 1993). Also Dominique Godineau, "Masculine and Feminine Political Practice during the French Revolution, 1793-Year III," in *Women and Politics in the Age of the Democratic Revolution.*

113. Levy and Applewhite, "Women and Militant Citizenship," 85.

114. Cited in Darline Levy, Harriet Branson Applewhite, and Mary Durham Johnson, eds. and trans., *Women in Revolutionary Paris 1789–1795: Selected Documents* (Urbana: University of Illinois Press, 1979), 72–74, emphasis mine.

115. Godineau, "Masculine and Feminine Political Practice," 65. According to Godineau, "the originality of women's political practices seems . . . to rest far more on the importance of the idea of popular sovereignty than on the subsistence question. The *sans-culottes,* men as well as women, had a very concrete idea of this concept: the Sovereign People deliberating in its section assemblies, bearing arms, sitting on the Revolutionary Tribunal. . . . Women of the people identified themselves as belonging to the Sovereign, and no one thought of denying that the Sovereign People was composed of both citizens and *citoyennes.* But women were legally excluded from the body politic and possessed none of the attributes of sovereignty (voting rights, the right to deliberate in the general assemblies of the sections, the right to organize in an armed body and sit on the Revolutionary Tribunal, etc.). Their status was ambiguous—that of *citoyennes* without citizenship. On the other hand, in case of popular insurrection, when the sovereign People made its voice audible to mandatories whom it judged unfaithful, women

constituted an integral portion of the Sovereign—although such insurrection was not included within the structures of *sans-culotte* organization. In this paradoxical situation,
women were fully recognized as members of the Sovereign only in times of insurrection, when the people attempted to reclaim its rights, rights that women, as women, did
not enjoy" (68).

116. Levy, Applewhite, and Johnson, *Women in Revolutionary Paris,* 161–65.

117. Levy, Applewhite, and Johnson, *Women in Revolutionary Paris,* 205.

118. Levy, Applewhite, and Johnson, *Women in Revolutionary Paris,* 213.

119. Rousseau, *First Discourse,* 221, emphasis mine.

120. Levy and Applewhite, "Women and Militant Citizenship" and "Women, Radicalization and the Fall of the French Monarchy."

121. Judith Butler, "Imitation and Gender Insubordination," in *The Lesbian and Gay
Studies Reader,* ed. Henry Abelove, Michèle Aina Barale, and David M. Halperin (New
York: Routledge, 1993), 313.

122. Marjorie Garber, *Vested Interests: Cross-Dressing and Cultural Anxiety* (New
York: HarperCollins, 1992), 16.

123. Garber, *Vested Interests,* 10.

Chapter 4

The Civic Rituals of the
American Citizen-Soldier

For years political theorists have presented competing narratives about the nature of American history and politics. How should we understand our heritage? Has there been a "liberal consensus" in American politics, as Louis Hartz argues?[1] Or are the "republican revisionists" and communitarians correct in stressing the vital importance of civic republicanism in early America?[2] Is America a nation of self-interested individuals or of virtuous citizens concerned about the public good? Is the "public good" simply the sum of individual goods? Or is it something that arises out of the transcendence of individual interests? Who are we as American citizens?

These questions are not just academic. The story we tell about our past directly affects the kind of politics we make in the present. As Benjamin R. Barber argues, "the historian uncovers in history the justification he seeks for his own time."[3] This is not to say that the historian or political theorist simply subordinates standards of scholarship to a particular political agenda. Instead, every scholar of American history and politics attempts to make sense out of the "confusing compound of political traditions and civic rhetorics" that constitutes the era of the American Founding. In so doing, each ends up constructing a particular narrative about our American heritage. Thus, political historiography necessarily involves the interpretation and evaluation of facts: "Every interpretation is admissible, none sovereign. . . . American history would seem to be a fable not agreed upon, a fable told and retold by historians with distinctive visions not of the past or even of the present, but of a future that might be."[4] How we narrate the story of American history directly affects the possibilities we see for American citizenship today.

On the one hand, stressing the Lockean origins of American politics produces a liberal picture of contemporary American politics. If we begin, as liberalism does, with the assumption of "atomistic social freedom," we will end up with an inadequate account of citizenship. While liberalism can justify the

proliferation of "rights" that serve to protect the freedom of atomistic individuals from the infringement of others, it has a harder time articulating a notion
of citizenship, which entails civic obligations to other members of the community.[5] On the other hand, remembering the centrality of civic republican ideals
to the creation of the America republic helps legitimize contemporary attempts
to resuscitate the ideals of participatory citizenship, civic virtue, and government aimed at the common good.

The Founding of the Citizen-Soldier Tradition in America

Civic republicanism profoundly shaped the ways in which early Americans con-
ceptualized politics. In fact, according to J. R. Pole, a republican consensus existed among early Americans. On both sides of the debate over the Constitution,
he tells us, "the men of Philadelphia enjoyed the advantage of a pervasive consensus of principles." No one argued for monarchy or aristocracy because "all
subscribed in one form or another" to "republican principles."[6] And this is clear
from the writings of both federalists and antifederalists, whose arguments are
peppered with appeals for "liberty" and "virtue" and against "tyranny" and "cor-
ruption"—all concepts central to the historic discourse of civic republicanism—
as well as explicit and/or implicit references to the Roman Republic, Machiavelli,
Montesquieu, and Rousseau.[7] Early Americans understood themselves as heirs
of what J. G. A. Pocock has termed "the Atlantic Republican Tradition," which
began with the work of Niccolo Machiavelli. Gordon S. Wood concurs with this
interpretation: "For Americans the mid-eighteenth century was truly a neoclassical age."[8] This neoclassical understanding of politics is evidenced by the frequent
use of the names of ancient republican heroes by colonial and early American citizens and by the "classical references and allusions" that ran "through much of
the colonists' writings, both public and private." In fact, "it was a rare newspaper
essayist who did not use a Greek or Latin phrase to enhance an argument or em-
bellish a point and who did not employ a classical signature."[9]

The early American allegiance to the ideals of civic republicanism included
a commitment to the ideal of the Citizen-Soldier. For example, the most explicit
advocates of civic republicanism during the founding, the antifederalists, emphasized the traditional republican fear of a standing army. In order to ensure
the protection of republican liberty and self-rule from tyranny, the antifederalists embraced the Citizen-Soldier tradition as previously articulated by Machiavelli and Rousseau. One civic republican, known only as "a Farmer," ex-
plained that "*a standing army,* in our present unsettled circumstances . . . would
wield us into despotism in a moment, and we have surely had throat-cutting
enough in our day."[10] A citizen called "Brutus" argued that "keeping up a
standing army would be in the highest degree dangerous to the liberty and
happiness of the community," and so "the general government ought not have

authority to do it; for no government should be empowered to do that which if done, would tend to destroy public liberty."[11] Patrick Henry added that without arms, the people would not be able to "punish tyrants."[12] In short, the antifederalists maintained that "a free republic will never keep a standing army to execute its laws. It must depend upon the support of its citizens."[13] In so doing, they laid claim to the ideal of the Citizen-Soldier.

In keeping with the civic republican tradition, early Americans viewed the polity as reconciling the common good with individual interests. So while contemporary communitarian scholars such as Michael Sandel and Gordon Wood have a point when they stress that the "Bill of Rights," in its origins and prior to the Fourteenth Amendment, aimed at securing the rights of the states vis-à-vis the federal government, rather than ensuring the rights of the individual vis-à-vis the community,[14] at the same time we must also recognize that early Americans viewed the "common good" of the community as *including* the interests of each individual citizen. As Wood himself concedes, "since everyone in the community was linked organically to everyone else, what was good for the whole community was ultimately good for all the parts."[15] Thus, the common good was not seen as standing opposed to individual interests.

The Second Amendment grew out of this synthetic relationship between the individual and the community. In the Second Amendment "the armed citizen and the militia existed as distinct, yet dynamically interrelated elements within American thought; it was perfectly reasonable to provide for both within the same amendment to the Constitution."[16] In other words, like the First Amendment, the Second Amendment constitutionally protects the ability of individuals to engage in the civic and martial practices constitutive of republican citizenship. Early American civic republicans did not understand the individual and the community as oppositional entities.

The Citizen-Soldier tradition constitutes the overall context in which early Americans imagined their civil-military relations. The Second Amendment and the Militia Act of 1792 institutionalized this tradition.[17] While the Second Amendment constitutionally protected the right of citizens to form militias, the Militia Act fixed the *principle* of a universal military obligation in the statutory law of the new government. It required the enrollment of every free, white, able-bodied male citizen between 18 and 45 in the "unorganized" or "common" militia of his state and required each man to provide his own weapons.[18]

In *The Soldier and the State* Samuel P. Huntington delineates "three strands of American militarism," two of which relate directly to the Citizen-Soldier tradition. "The popular strand" explicitly includes the civic militia. However, the "technicism" strand also grows out of the Citizen-Soldier tradition. As Huntington explains, within the "technicism" strand

> the officer was expert in one of several technical specialties, competence in which separated him from other officers trained in different specialties and at the same

time fostered close bonds with civilians practicing his specialty outside the military forces. The officer corps, in other words, was divided into subgroups, some more important than the rest, but each likely to be more closely tied with a segment of civilian society than with other segments of the corps.[19]

In other words, "technicism" reinforces the blurring of the distinctions between the military and civil society and so hinders the emergence of a distinctive military professionalism, one of the goals of the Citizen-Soldier tradition. Thomas Jefferson, the key figure within the "technicism" school, condemned "the distinction 'between the civil and military, which it is for the happiness of both to obliterate.' "[20] Jefferson favored a strong civic militia system with universal military obligation and wanted to make military training a key part of college education. His founding of West Point in 1802 profoundly influenced American military education.[21] My interpretation of the "technicism" strand as part of the Citizen-Soldier tradition makes sense of Jefferson's simultaneous advocacy of the Citizen-Soldier ideal and his founding of West Point.[22]

American Civil Society and the *Citizenship of Civic Practices*

In the eighteenth and nineteenth centuries, a strong civil society existed in America in which individuals became citizens as they engaged together in civic and martial practices.[23] In *Democracy in America,* Alexis de Tocqueville comments on the role civic practices traditionally played in creating citizens who are committed to the republic:

> The native of New England is attached to his township because it is independent and free: *this cooperation in its affairs insures his attachment to its interest;* the well-being it affords him secures his affection; and its welfare is the aim of his ambition and of his future exertions.[24]

The concern with the common good that constitutes citizenship is constructed through the practices of self-government. "Patriotism," he argues, "is strengthened by ritual observance," by engagement in the practices of citizenship required in a system that "divides local authority among so many citizens, [and] does not scruple to multiply the functions of the town officers."[25] New England is "thoroughly democratic and republican."[26]

Tocqueville argues that, absent participation in self-rule, republican citizenship cannot exist. In making this argument he contrasts the civic action of the American citizen to the apathy of the European subject. "There are countries in Europe," he explains, "where the natives consider themselves as a kind of settlers, indifferent to the fate of the spot which they inhabit." Their indifference arises from the fact that "the greatest changes are affected there without their concurrence." Because the European subject views government as "unconnected with himself," it does "not concern him." Even if "his own safety or that of his children is at last endangered," he does not act but instead "folds his arms, and wait[s] till the whole nation comes to his aid. . . . When a nation has arrived at this state, it must either change its customs or its laws, or perish; for the source of public virtues is dried up; and though it may contain subjects, it has no citizens."[27] Without civic participation, patriotism and civic virtue cannot exist, and so neither can republican citizenship.

In opposition to the European subject, Tocqueville argues, the American citizen does not look to the state to solve his problems. Instead, American citizens solve their own problems through the process of civic deliberation and civic action:

> When a private individual meditates an undertaking, however directly connected it may be to the welfare of society, he never thinks of soliciting the cooperation of the government; but he publishes his plan, offers to execute it, courts the assistance of other individuals, and struggles manfully against all obstacles.[28]

For example,

> if a stoppage occurs in a thoroughfare, and the circulation of vehicles is hindered, the neighbors immediately form themselves into a deliberative body; and this extemporaneous assembly gives rise to an executive power, which remedies the inconvenience before anybody has thought of recurring to a pre-existing authority superior to that of the persons immediately concerned.[29]

In fact, civic deliberation is such a central part of American life that "no sooner do you set foot upon American ground, than you are stunned by a kind of tumult; a confused clamor is heard on every side; and a thousand simultaneous voices demand the satisfaction of their social wants."[30] The interaction of diverse perspectives is productive: "He is canvassed by a multitude of applicants, and, in seeking to deceive him in a thousand ways, they really enlighten him."[31] In fact, "the cares of politics engross a prominent place in the occupations of a citizen in the United States."[32] In this way, early American citizens exercised "the most natural privilege of man, [which] next to the right of acting for himself, is that of combining his exertions with those of his fellow-creatures, and acting in common with them."[33] In Tocqueville's estimation at least, a strong civil society, a key component of civic republicanism, existed during the early years of American history.

Tocqueville, however, was not uncritical of American civil society. First of all, he feared that the existence of (what he saw as) equality of condition could eventually lead to radical individualism and the erosion of society. As he explains, democratic society lacks the hierarchical organization that in aristocratic society ensures bonds of obligation and dependency between men.

Aristocracy had made a chain of all the members of the community, from the peas-
ant to the king: democracy breaks that chain, and severs every link. . . . Thus, not
only does democracy make every man forget his ancestors, but it hides his de-
scendants and separates his contemporaries from him; it throws him back forever
upon himself alone, and threatens in the end to confine him entirely within the solitude of his own heart.[34]

Thrown back on himself alone, democratic man develops a sense of individualism

which disposes [him] to sever himself from the mass of his fellows, and to draw
apart with his family and his friends; so that, after he has thus formed a little circle of his own, he willingly leaves society at large to itself. . . . [I]ndividualism, at
first, only saps the virtues of public life; but, in the long run, it attacks and destroys
all others, and is at length absorbed in downright selfishness.[35]

In a nonhierarchical, democratic society, "the bond of human affection is extended, but it is relaxed."[36] Each man becomes concerned only with the interests of himself and his close circle of family and friends.

For Tocqueville only participation in civic practices can counteract the frag-
mentation of democratic society. "It is difficult to draw a man out of his own
circle to interest him in the destiny of the state," Tocqueville reasons, "because
he does not clearly understand what influence the destiny of the state can have
upon his own lot." However, once a man takes action at a local level to achieve
his own self-interest, he begins to see the connection between his private interest and the public good.

Thus, far more may be done by intrusting to the citizens the administration of mi-
nor affairs than by surrendering to them the control of important ones, towards interesting them in the public welfare, and convincing them that they constantly
stand in need one of another in order to provide for it.[37]

Furthermore, through engagement in civic practices at the local level the man
becomes the citizen who cares for the public good. The need for common action "perpetually brings men together, and forces them to help one another, in
spite of the propensities which sever them." As a result "a great number of citizens [learn] to value the affection of their neighbors and of their kindred."[38]
While men "attend to the interests of the public, first by necessity, afterwards
[they do it] by choice."[39] Civic participation creates citizens and instills in them
a sense of patriotism, fraternity, and civic virtue.

Tocqueville's second fear was that democratic civil society could produce a
"tyranny of the majority." Because all power in a democracy derives from the
people, he argues, no standard of judgment exists outside the decisions of the
majority: "When an individual or a party is wronged in the United States, to

whom can he apply for redress?" All political institutions represent the majority in one way or another.[40] Absolute sovereignty of the community means that there are no independent standards by which community decisions can be judged. Consequently, the interests of minority groups could easily be overrun by the decisions of the majority, and there would be no appeal. This overwhelming power of the majority could lead to tyranny: "Unlimited power"— even of the people in a democracy—"is in itself a bad and dangerous thing," Tocqueville cautions, because "human beings are not competent to exercise it with discretion" as is God.[41]

Carrying his critique even further, Tocqueville argues that the power of the majority could also hinder the very formation of dissenting opinions, thus truncating political debate. That is to say, in America "the majority possesses a power which is physical and moral at the same time, which acts upon the will as much as upon the actions, and represses not only all contest, but all controversy." Tocqueville insists the he knows "of no country in which there is so little independence of mind and real freedom of discussion as in America."[42] While the process of civic deliberation Tocqueville praises can produce a synthesis of individual interests into a common good, it can also exert a pressure on participants to conform to the majority opinion.

Nevertheless, despite this possibility, the American version of civic deliberation emphasized by Tocqueville allows room for a lot more diversity and dissension than does Rousseau's version. As we saw in chapter 3, Rousseau emphasized a unitary *general will* constructed through civic practices that did not allow for dissension. In contrast to this, Tocqueville stresses the important role played by disagreement and debate within the American context. In America "a thousand simultaneous voices" engage in civic deliberation in order to "demand the satisfaction of their social wants."[43] Thus, this American version of the *citizenship of civic practices* does not require the elimination of divisive issues from political deliberations. Instead, it presents a model of civic deliberation that aims to forge a common good out of a diversity of different opinions.[44]

The Martial Rituals of the Citizen-Soldier

However, although Tocqueville discusses at length how the process of civic participation contributed to the construction of citizenship in early America— the first half of the Citizen-Soldier ideal—he overlooks the martial practices that also played a key role in the constitution of republican citizenship during this time period. That is, it was precisely because of the greater degree of dissension present within American politics that the martial practices of the Citizen-Soldier tradition formed such a vitally important function in unifying a potentially individualistic citizenry. In American politics as in Machiavellian and Rousseauian politics, the *citizenship of civic practices* characteristic of the

Citizen-Soldier tradition requires the engagement in *both* civic *and* martial practices. Both halves of this ideal are equally important.

Consequently, the strong tradition of civic republicanism that existed in America through the nineteenth century featured the Citizen-Soldier ideal as a central category.[45] Although not always actualized in practice, the Citizen-Soldier constituted a cultural and political ideal that was more important for the production of masculine citizens than for actual military effectiveness. More specifically, the Citizen-Soldier ideal functioned by prescribing a set of martial *rituals* in which male individuals must engage if they were to become masculine republican citizen-soldiers. The concept of ritual is important here because, as Catherine Bell has theorized, rituals function particularly well in unifying a diverse group of individuals into a coherent whole. "Solidarity," she explains, derives "not from the formulation and communication of coherent beliefs held in common by participants but from the *activities* of ritual per se."[46] Thus, especially in a context in which "a thousand" competing voices exist, participation in a set of common rituals allows for the creation of solidarity. Thus, rituals create community without sacrificing individuality.

Missing the important *political* role played by martial rituals, many military historians have emphasized the military inadequacies of the civic militia. For example, one scholar argues that while the civic militia had formed the basis of colonial defense, during the American Revolution—the supposedly quintessential example of citizen-soldiery at its finest—the militia actually "proved ineffective."[47] Another stresses that the early American "glorification of the militia—the myth of the citizen-soldier"—

> developed in the face of considerable evidence of its inadequacies. . . . Time and experience revealed the distance between ideal and reality. In their ideology, the regular army was "only ancillary to the revolution." The reality was quite different; a regular force had been essential to victory.[48]

In still another historian's words,

> the impulse to glorify the revolutionary effort led to exaggerated claims of success and helps to explain the significance accorded the militia by Americans in the 1780's. The popular interpretation of victory in the Revolution ignored the role played by the regular army and reinstated the people's militia as the vital pillar of American virtue and essential to the preservation of the nation's unique republican character.[49]

So even though "by 1790 the militia had lost its status as a viable military institution," he tells us, it continued to exist as a symbol of republican liberties.[50]

Again, what these scholars fail to recognize is that the myth of the Citizen-Soldier functioned as a vitally important cultural and political ideal in eighteenth- and nineteenth-century America because it demanded that a diversity of male individuals participate in the martial rituals that would constitute them as a fraternity of masculine republican citizens. Consider this contemptuous portrayal of early American citizen-soldiers:

> The men thus enrolled *en masse* by statute were far more interested in somehow demonstrating a constitutional right to bear arms than they were concerned about ever using those arms for national defense or anything else. For the purpose of effecting this demonstration, as well as to engage in strutting display and to gain pseudosocial recognition among some of their equally shortsighted neighbors, the militiamen in various villages and towns were periodically gathered together in company-sized groups for drill and training, according to law.[51]

The argument continues:

> Almost from the beginning and thereafter throughout the next 150 years, until abolished when the National Guard was established by the Dick Act of 1903, American *nonvolunteer* militia units were essentially social in nature, incapable of serious military functioning, and more avoided by the regular army than welcomed.[52]

What such arguments overlook is that the civic militia fulfilled an important social and *political* function in republican America, because its rituals played an important role in the constitution of patriotism, fraternity, civic virtue, citizenship, and masculinity. Participating together in martial rituals, white male individuals came to constitute a fraternity of manly citizens who loved their community and so were willing to think of the common good rather than just their own individuals interests.

For example, the practices of the civic militia traditionally played a key role in the creation of fraternity in early American politics. This role was so important that when the common militia tradition of universal and obligatory service deteriorated during the early nineteenth century, the volunteer militia tradition "emerged, partially filling the military void."[53] These "volunteer units satisfied the sort of public demand that sports were to fulfill later on" and kept "the martial spirit alive in regions more and more remote from immediate danger."[54] Individuals became a fraternity of manly citizen-soldiers as they participated in the practices of the civic militia.

The civic and martial practices embodied in the figure of the Citizen-Soldier animated American politics during the nineteenth century and were central to what Michael E. McGerr calls "spectacular politics," which began in the 1820s and 1830s and reached their height during the American third-party system (1850s to early 1890s). During this period the political parties regularly sponsored civic festivals that had martial practices at their center:

> For presidential contests, and occasionally those state contests charged with national issues, Democrats and Republicans used the traditional elements of

political display more extensively than ever before and added *the uniformed marching company* to create an intense, enveloping partisan experience. . . . Men may *have paraded in uniform* occasionally during Jacksonian elections, but *the military company became the trademark of spectacle in the third party system.*[55]

In nineteenth-century America, political parties sponsored civic festivals like the ones Rousseau advocated.

Moreover, John Mahon stresses, in his *History of the Militia and the National Guard,* that military uniforms played a key role in stimulating participation in civic festivals:

> In some of the people, the martial spirit combined with a *love of colorful uniforms,* military ceremonials, and martial music was ever present. Such props appealed to the vanity of many persons and to the noble instinct of others, and through them citizens of all sorts sought to escape being ordinary. . . . *The opportunity to wear a uniform attracted some men, the gaudier the uniform the better.* . . . As *uniforms enlarged the self-respect of the wearers,* so also did they stir those persons who saw them. . . . *Festivals would have been drab without the volunteer militia, the units of which were easy to involve in public appearances.* . . . They also conducted target shoots, and marched with much ceremony to visit neighboring units. The encampments occasioned by these visits involved themselves and the host communities in gargantuan feasts, much fancy drill, and sham battles.[56]

Thus, the martial practices of the civic militia served a vital function in nineteenth-century America, because they helped stimulate the civic participation through which masculine citizens would be constituted.

Civic and martial practices were interconnected during the era of strong parties. "Spectacular campaigns," McGerr tells us,

> mingled the intellectual stimulation of an open-air, hour-long oration on the tariff with the military nostalgia of the uniformed company. Partisan display combined the exertion of long marches with the delights of a fireworks show. Transforming communities into partisan tableaux, spectacle fused martial dreams, intellectual endeavor, leisure enjoyment, and hard labor in the service of politics.[57]

Both halves of the Citizen-Soldier ideal, participatory citizenship *and* service in the militias, were important during this period. "Together the clubs and [marching] companies created a partisan spectacle that engulfed Northern communities for the three months before election day."[58] Individuals became citizens as they engaged in the civic practices and martial rituals prescribed by the ideal of the Citizen-Soldier.

Participation in the political process is important to the constitution of citizens within the civic republican tradition because it requires that individuals work together to govern themselves for the common good, rather than simply pursuing their own interests. From McGerr's description of nineteenth-century civil society, we can discern the existence of high levels of civic deliberation. That is, we can assume, given the centrality of party newspapers, partisan issues, and political speeches in this era of "spectacular politics," that individuals throughout the community actively engaged in civic deliberation in both formal and informal ways. Certainly political speeches were a central part of "spectacular politics": "Unable to find seats inside [to hear visiting party heroes speak], thousands of people often stood in the streets to hear orators speak from makeshift platforms."[59] Despite the popularity of "party heroes" during this period, campaigns were waged on the basis of issues, not personalities; the actual presidential candidates played a limited role in their own campaigns. All candidates were expected to do was accept the nomination and endorse the party platform. In fact, campaigning vigorously on one's own behalf was considered to be in poor taste.

My theorization of the *citizenship of civic practices*—that individuals *become* citizens only as they participate in civic and martial practices—augments McGerr's analysis of why there was a strong positive correlation between "spectacular politics" and voting. In answering this question McGerr argues that, in the first place, political parties stimulated voting by drawing people into the political process. More importantly, however,

> the significance of political spectacle and party journalism lay not so much in their effect on voting at this or that presidential election as in their *influence on the habits of the generations of men* voting at all elections, local as well as national, in the nineteenth century.[60]

Spectacular campaigns "captivated" and "initiated" young boys into politics.[61] In other words, engagement in civic and martial practices actually *constituted* them as citizens.

That is to say, partisanship became an important aspect of civic identity:

> Through participation in spectacular campaigns, Northerners revealed their belief not merely in the legitimacy of party commitment, but also in the necessity of demonstrating that commitment in public before the community. Like the party press, political spectacle made partisanship appear an *integral element of men's identity* and outlook.[62]

More pointedly, partisanship was central to civic identity because it caused men to engage in civic action and only through that action could they *become* citizens. In other words, *citizen* was not a prepolitical identity in nineteenth-century America, but rather was an identity constructed through civic practice.

The rituals required by the Citizen-Soldier ideal and enacted by the marching companies played a key role in the constitution of masculine republican citizenship. Focusing on nationalism, Lauren Berlant argues that "*participation* in national celebration connects the citizen to a collective subjectivity constituted by *synchronous participation in the same national rituals,* the same discursive system."[63] Similarly, the martial rituals and civic festivals characteristic of the Citizen-Soldier tradition functioned to connect individuals to a collective subjectivity—that is, to their republican communities. Citizenship was actually constructed through synchronous participation in civic and martial rituals. However, while Berlant focuses on the role the Statue of Liberty played in the creation of American nationalism at the end of the nineteenth century, the ideal of the Citizen-Soldier required participation in the *states'* militias and thus helped constitute not *national* citizenship but citizenship that was rooted in local communities within particular states. Indeed, the emergence of American nationalism actually corresponded to a decline in the centrality of the Citizen-Soldier ideal to American politics.

Many scholars dismiss the importance of civic practices in nineteenth-century American politics. For example, attacking valorizations of the highly attended Lincoln-Douglas debates as examples of civic participation and deliberation par excellence, Michael Schudson maintains that while "it is true" that many people attended the debates, "it is not at all apparent what in those debates they attended to. It is true that they participated, but it is not clear that they were 'interested in issues of transcendent importance.' " Political campaigns in the nineteenth century, Schudson argues, were more "religious revivals and popular entertainments than settings for rational-critical discussion."[64] Recalling his own experience in the 1960s, Schudson insists that "there is a big difference between attending a rally and actually listening to the speeches." He concludes that "the idea that a public sphere of rational-critical discourse flourished in the eighteenth or early nineteenth century, at least in the American instance, is an inadequate, if not incoherent, notion."[65] For Schudson, only rational discourse counts as politics.

While "a public sphere of rational-critical discourse" might not have ever existed in America, Schudson is missing the important point: Nineteenth-century America was characterized by a vibrant civil society in which individuals became citizens as they engaged together in civic and martial practices. It was through engagement in these practices that individuals developed the *passionate* connection to their communities—the feelings of patriotism, fraternity, and civic virtue—that underwrote their interest in self-government. Thus, the important question is not whether people actually listened to the arguments made in political debates and deliberated about the issues—although, given the context, it is hard to believe that nineteenth-century Americans did not do this. What matters more is the fact that individuals were being transformed into republican citizens, as they attended rallies, gathered in the streets, sang songs, picnicked, wore uniforms in strutting display, and demonstrated their constitutional right to bear arms in celebratory parades.

Furthermore, Schudson, like many other "public sphere" advocates, misunderstands what constitutes citizenship within the civic republican tradition. In a direct attack on McGerr's scholarship, Schudson argues that "spectacular politics" was

more a communal ritual than an act of individual or group involvement in rational-critical discussion. . . . It was organized much less with the rational choice of
the individual voter in mind. *The voter, in a sense, was not conceived of as an individual but as an entity enveloped in and defined by social circumstance and party
affiliation.*[66]

And this is exactly the point: The civic republican citizen-soldier was precisely an enveloped entity constituted through engagement in civic and martial practices.

As I have been arguing, the civic republican tradition does not conceptualize
citizens as prepolitical entities—rational individual voters—who then choose
whether or not to engage in political discourse or patriotic martial rituals. Instead, the individual actually *becomes* a citizen as he participates in "communal
ritual." In other words, for civic republicanism, citizenship should be understood
as a set of practices that eventually produce a civic identity, rather than as a prepolitical category established through residence within a particular bounded territory *(ius solis)*—or through having a particular race or ethnicity *(ius sanguinis)*. While those *allowed* to participate in the practices constitutive of
republican citizenship may in *fact* have the same race or ethnicity, civic republicanism does not restrict citizenship in this way as a matter of *principle*.

In sum, up through the end of the nineteenth century, the Citizen-Soldier
ideal played a central role in the constitution of masculine republican citizenship in America. This ideal represented the linkage between participatory citizenship and military service—at least at the level of ideology. The ideal of the
Citizen-Soldier led male individuals to engage in the civic and martial practices
through which they would be transformed into masculine citizen-soldiers.
These practices produced feelings of patriotism, fraternity, and civic virtue as
well as a common civic identity—all of which provide the necessary foundation for republican self-government aimed at the common good. Male individuals became masculine republican citizen-soldiers as they participated in martial rituals and civic festivities organized by the political parties.

Policing the Borders of the Civic Republic:
The Vices of the Citizen-Soldier Tradition

Vices as well as virtues characterize the Citizen-Soldier tradition. While civic
and martial practices create citizen-soldiers and instill in them patriotism, fraternity, civic virtue, and a common civic identity, these same practices can also
yield chauvinism, racism, violence, and homogeneity. In nineteenth-century
America, participation in civic and martial practices did indeed create virtues
that underwrote participatory citizenship. But participatory citizenship also
produced a set of undemocratic vices. The citizen-soldier identity was forged

in opposition to denigrated "others." The question for democratic theorists is whether we can augment the virtues of the Citizen-Soldier ideal while downplaying its vices.

The Citizen-Soldier tradition that was so essential to the constitution of republican citizenship in America produces a package of interconnected virtues and vices. Richard Moser agrees on the importance of the Citizen-Soldier tradition to American political culture. However, in his study of the role of soldiering in American history and historiography, Moser argues that there are actually *two* important and conflicting soldier ideals in American culture: the Citizen-Soldier and the Fighter. Moser concludes that "the American soldier ideal is defined by the tension between [these] two opposing historical traditions, each with its own meanings and myths."[67] But while Moser provides important insights into the role of the Citizen-Soldier ideal in American culture and on the constitution of *armed masculinity* with the U.S. military—a topic we will discuss more fully in chapter 6—I disagree with his stark separation of the Citizen-Soldier and the Fighter. In essence, Moser locates all of the virtues of the Citizen-Soldier tradition in the former figure and all the vices in the latter. But as we shall soon see, the practices of the Citizen-Soldier tradition itself produce both virtues and vices.

Indeed, the virtues and vices go hand in hand. For example, individuals became republican citizen-soldiers as they waged war on Native Americans. In fact, one of the earliest and most important reasons for maintaining a militia system was to engage in a war to contain Native Americans that in time became genocide.[68] During the Jacksonian period of American democracy, removal of the eastern Indians to areas west of the Mississippi was a major project. The Black Hawk War during the early 1830s "showed the citizen soldiery at its worst." For example, "many of the short-term irregulars considered these redmen to be animals, much lower on the life scale than man. They wanted this animal out of the way and welcomed the chance to kill it." Hatred of the Sauk and Fox Indians "grew among the citizen soldiery. . . . Some of them, finding a few Indian women furrowing in the river bank to hide, shot them and especially relished watching them jerk as they died." The Indians "became the active enemy of the militia, both standing and volunteer, during the decades following the War of 1812."[69] White male Americans constructed their civic identity through violent martial practices that annihilated Native American populations.

American citizen-soldiers likewise constituted their identity in opposition to the Mexicans, as they patrolled America's southern border. During the Mexican War in the 1840s, many of the volunteer citizen-soldiers "considered the Roman Catholic Mexican peasants as being on a low rung of the life ladder, no higher than the North American Indians." More than once the regular army soldiers intervened to save "native Mexicans from rape, pillage, and death at the hands of the volunteers" of the militias. One regular soldier wrote that he saw the volunteer militiamen "fighting over their victims like dogs, and the place resounded with horrid oaths and the groans and shrieks of the raped."[70] Once again we see civic identity constructed in opposition to denigrated "others."

In his history of the civic militia, Mahon argues that the amateurism and disorderliness of the citizen-soldiers actually increased the cruelty of both the annihilation of the Native Americans and the conquest of the Mexicans.[71] Driven by the passions of patriotism, citizen-soldiers often acted without restraint. Despite this reality, the American people saw the performance of the citizen-soldiers in the Mexican War as a great triumph, "a magnificent conquest. . . . The war had shown that the martial spirit was very much alive in the nation, a spirit essential to the mood of Manifest Destiny abroad in the land." The citizen-soldiers "vindicated the national honor"—but at the expense of the Mexican people.[72]

Not only did the citizen-soldiers define themselves in violent opposition to denigrated "others," but they also functioned best as a militia unit when membership was most homogeneous. During the seventeenth and early eighteenth centuries, the militia were "most effective in New England . . . perhaps [in part because of its] considerable ethnic and cultural homogeneity."[73] Since citizenship was linked to compulsory service in the militia, as American citizenship was expanded in the early nineteenth century to include poor and working-class white men, the militias became increasingly heterogeneous. Along with this increasing diversity, we see the emergence of a movement in opposition to compulsory participation in the militias. Urban middle- and upper-class men in the North began to oppose militia duty, while the urban press began to ridicule the newly inclusive militia, calling it " 'rabble' or 'scarecrow militia,' engaged in a parody of military drill, many without weapons or uniforms. The militia-training days, critics argued, had degenerated into loafing and insobriety."[74] During the period of Jacksonian democracy, the "unorganized" militias made up of all citizens were replaced by volunteer militia companies "composed of like-minded men from the same town or ethnic group who enjoyed the pomp and camaraderie of a military brotherhood." These new militia units called themselves the "National Guard," partially in order "to distinguish themselves from the disreputable general or common militia."[75]

The South, on the other hand, continued to support mandatory militia training for all white men, because the southern militias played a vital role in the maintenance of white supremacy, patrolling the southern plantations and suppressing slave insurrections.[76] Moreover, in addition to militia duty, "every able-bodied man had the obligation periodically to join in nocturnal neighborhood [slave] patrols," conducted by the "extralegal and semimilitary organizations" that emerged to "supplement the state militias and sheriffs' posses which the law sanctioned."[77] Service in the slave patrol was considered an extension of the white man's civic militia duty.[78]

With millions of Afro-Americans in the region, most southern whites viewed the local militia and slave patrol as essential instruments of race control. Adroitly, southern leaders emphasized that the militia encouraged martial skills and virtues, highly prized among rural residents, and also forged a bond among white males.[79]

The militia was "the corporate embodiment of white male political fraternity."[80] This was true to some extent even in the North, where the Fugitive Slave Law of 1850 brought work to the militia units, which were involved in roundups of runaway slaves.[81]

Militia service in the South bolstered simultaneously both fraternity and racism. In fact, militia service played such an important role in the maintenance of white supremacy in the South that after the Civil War the Republican party denied the ex-Confederate states the right to form militias.[82] Bulwarks of white supremacy, the southern militias would not act to suppress widespread violence by whites against African Americans. One could say they saw violence as necessary for the common good. To the Republican party, "a militia composed in large part of the very white men who were engaged in lawlessness, or were sympathetic with it, seemed worse than useless."[83] However, because of the constitutional protection of the militias by the Second Amendment and because the regular army needed assistance in their occupation of the South, Congress reinstated the Southern militias but restricted membership to African Americans and white Unionists. Needless to say, the arming of African Americans in militia units was "intolerable to the southern whites."[84] The specter of the long-feared "race war" threatened in their minds. Southern whites viewed the African American citizen-soldiers as an "obnoxious" bunch of "swaggering bullies" and feared that the African American militias would become "a standing army of Negro soldiers."[85]

With African Americans in the official militias, white Southerners had no intention of abandoning their traditional martial practices. Vowing "to use unbridled violence if necessary," white men across the South began to form "white rifle companies" to replace the dissolved state militias.[86]

> Although they lacked official sanction, these companies had behind them the determination of the society to establish white supremacy at all costs. . . . White riflemen ambushed and killed black officers and white supporters of the Negro militia. These assassinations often took place in broad daylight with witnesses, but prosecutions were nonexistent.

As a consequence,

> the leadership that supported the black militia was either killed or intimidated. When the Democrats returned to power in state after state of the ex-Confederacy, they terminated the black militia, disarmed the blacks, and excluded them from any role in the militia.[87]

The white rifle companies played a key role in overthrowing Reconstruction.[88]

Moreover, the Ku Klux Klan emerged as a part of this white martial reaction to the enfranchisement of African Americans.[89] This terrorist organization historically grew out of the *practices* of the antebellum Southern militias that routinely enforced white supremacy, and thus it could be interpreted as an example of the Citizen-Soldier tradition at it most vicious. At the same time, however, it is important to note that as an antidemocratic organization, the KKK directly violates the *political theory* of the Citizen-Soldier, which, as I have argued throughout the book, contains within it a commitment to a set of universalizable humanist principles, including liberty, equality, and the rule of law. In addition, although the KKK believes in an active form of participation, it rejects the form of citizenship historically characteristic of the Citizen-Soldier tradition, the *citizenship of civic practices,* in favor of the *citizenship of blood* (*ius sanguinis*), which bases citizenship not on participation but on common bloodlines.

Issues of ethnicity as well as of race also affected the development of the civic militias during the last decades of the nineteenth century. That is to say, an increase of and changes in immigration over the latter half of the nineteenth century contributed to the changing composition of a formerly white, Anglo-Saxon, Protestant America. With the immigration of large numbers of Irish Roman Catholics in the 1840s, we see the beginning of a shift in immigration patterns that resulted in increased numbers of Jews and Catholics in the American populace and contributed to the growth of cities, which came to be seen as centers of discontent that "seemed to threaten an America that had been dominated by a homogeneous rural or small-town Anglo-Saxon Protestant culture."[90] By 1893 the majority of immigrants would be arriving from southern and eastern Europe (Italy, Austria-Hungary, Russia, Poland, Greece, and the Balkans) and from Asia (China and Japan), rather than from northern and western Europe (Great Britain, Ireland, Germany, and Scandinavia) as had been the case prior to the 1890s. The fact that many of these newcomers were working class created further anxiety. As the heterogeneity of the citizenry increased, so did discrimination and bigotry of all kinds: racism, nativism, xenophobia, anti-Semitism, anti-Catholicism, and fear of social democracy.[91]

These changes affected the militia in three ways. First of all, new immigrant groups, such as the Irish, began forming their own homogeneous, ethnic militias. Martial practices facilitated fraternal bonding and a common civic identity among participants. Part of civil society, these ethnic militia units were not organized by state governments, but rather were "created by social forces,"[92] that is, "by various ethnic and status groups, subsidized by the federal government, and fused with state and local party politics through ties of patronage."[93] New immigrants engaged in martial practices and in this way forged a sense of fraternity within their communities and a patriotic connection to the larger American republic.

Secondly, the proliferation of ethnic militia groups created fear in the hearts of WASP America and so fed the emergence of American nativism. "The growth of nativist and anti-Catholic sentiment . . . raised public fears about the loyalty and reliability of such ethnic National Guard units in a society increasingly divided along ethnic, religious, and class lines." Consequently,

in the 1850s, nativists seeking to create a "pure American" militia, demanded the exclusion of immigrants from the National Guard. In response, the governors of Massachusetts and Connecticut confiscated the weapons of all so-called "foreign militia" and ordered the dissolution of militia companies composed primarily of Irish Catholics.[94]

Engagement in martial practices produced a sense of fraternity and a common identity among new immigrants, and this created fear among white Americans who occupied a position of dominance.

Thirdly, the increasing diversity of class, race, and ethnicity in American society created fear on the part of the white, property-owning classes of the dark, urban proletariat gathering in the cities and led to calls for the building of a modern, professional army. As "swarms of immigrants entered the United States from southeastern Europe . . . [and] crowded into industrial cities," xenophobia and the fear of social democracy augmented each other. In the words of one general: "It is idle to close our eyes to the fact that there now exists in certain localities an element, mostly imported from abroad, fraught with danger to order and well-being unless firmly and wisely controlled."[95] In the discursive battle against the threat of social democracy, strikes were portrayed as "un-American," and the homogenized National Guard was called upon to protect private property from the increasingly heterogeneous people. In the words of one Guard leader, physical force was the only way to keep down the "savage elements of the society." In fact, the editor of the *National Guard* wrote that the laws the Guard upheld with its weapons were "enacted by the people before this fair country was overrun by the outcasts of Europe . . . villains from all parts of the Old World."[96] Doubt emerged about whether citizens' militias would protect private property from working people.

In fact, according to Stephen Skowronek, industrial strife was the main cause of calls for the replacement of the civic militias with a modern, American army.[97] America's unparalleled labor violence caused reformers to call for "a well-trained internal police force." In addition, America during this period was "an expanding commercial power [that,] no matter how favored geographically, required an international military capacity to protect its worldwide economic interests."[98]

The increasing heterogeneity of the American populous and the changes wrought by industrialization scared white, property-owning Americans and fueled an attack on both participatory citizenship and the civic militia—on both halves of the traditional Citizen-Soldier ideal. During the Progressive era, a transition occurred from the active, community-based *citizenship of civic practices* we have been exploring to a passive, individualistic, consumerist version of "citizenship" characteristic of contemporary America. Ultimately, the *citizenship of civic practices* gave way to the "professional politics paradigm"— the belief that we need "professionals" to govern for us—which accompanied the building of the "new American state." One of the key elements that enabled this transition was the attack on the mass political parties whose practices traditionally played a key role in transforming individuals into citizens. The changes begun during the latter part of the nineteenth century and consolidated in the early twentieth century ultimately put an end to both the participatory citizenship and the civic militia characteristic of the Citizen-Soldier tradition.[99]

Citizen-Soldiers and Working People

Between 1877 and 1900 a struggle ensued between those who favored the creation of a modern, professional military—namely, northern industrialists and military reformers—and those who stood with the long-standing American tradition of the Citizen-Soldier. The "new ideals of nationalism, expertise, and professionalism" ran up against "the deeply rooted traditions of volunteerism, federalism, and republicanism embodied in the institution of the state militia."[100] On the one hand, the Northern Republicans feared labor unrest, and did not trust the citizens' militias to protect the private property of the few. Consequently, Northern industrialists and their Republican Party favored the creation of a professional military they could trust to protect them from the threat of social democracy. On the other hand, Southern Democrats, who had returned to power during the 1870s, harbored "unveiled hatred for troops that had not only occupied the South but openly colluded with the Radicals to impose the Republican party and the Negro on southern politics."[101]Southern fears of a standing army were very real, and so they opposed the professionalization of the American military. Ultimately, the push for a modern, professionalized military was stymied by the long-standing American tradition of the Citizen-Soldier.[102]

When the Great Railroad Strike of 1877 erupted, citizen-soldiers were called upon to defend private property against working people—sometimes their own neighbors.

> During the early days of the strike, the old republican faith in citizen-soldiers appeared to threaten capitulation to the mob. Militia units faltered badly in actions at Baltimore, Philadelphia, and Pittsburgh. There were no organized militia units on hand when the strike reached Indiana and Missouri. Most fearful of all were reports of fraternization between militiamen and strikers in West Virginia, New

York, and Ohio. The specter of a militia collapse drew anxious appeals for federal troops.[103]

Despite the fears of industrialists, "in general, the Guardsmen on duty in 1877 carried out their orders even when their sympathies lay with the strikers."[104] Nevertheless, there was no guarantee that the citizen-soldiers would always act to repress their working neighbors for the benefit of the few.

Within the Citizen-Soldier tradition, the civic militia made up of all citizens stands ready to defend the people against tyranny. A professional army is feared because it can be used to advance the particular interests of the few against the common interests of the people. Thus it makes sense that as working people threatened the particular interests of industrial capital, calls for a professional army increased. For example, the corporate-sponsored *Chicago Tribune*

> favored the development of the regulars into a "national police force," because the *citizen-soldiers had proven too much a part of the society that now had to be controlled.* Unlike the amateur, the regular "has no politics, no affiliations, no connections with trade unions or corporations." The regulars were the independent strong arm of the state.[105]

In Samuel Gompers' words, "[S]tanding armies are always used to exercise tyranny over people."[106]

Industrial capitalists and their supporters feared that citizen-soldiers might very well side with social democracy. In the words of then Secretary of War McCrary:

> As our country increases in population and wealth, and as great cities become numerous, it must be clearly seen that there may be great danger of *uprisings of large masses of people for the redress of grievances,* real or fancied; and it is a well-known fact that such uprisings enlist in greater or lesser degree the sympathies of the communities in which they occur. This fact alone renders the militia unreliable in such an emergency.[107]

Even though the National Guard did indeed break up strikes and actually fired on strikers on several occasions, the possibility still existed that citizen-soldiers would not act against working people.[108] McCrary called for a cool, steady, and obedient professional military ready to stand firm against participatory popular sovereignty.[109]

Nevertheless, the Citizen-Soldier tradition was so rooted in American political culture that it could not be easily dislodged. Despite the needs of Northern industrialists and the desires of military reformers, as Skowronek demonstrates, entrenched traditions, institutions, and interests prevented the emergence of a modern, professional army at the end of the nineteenth century. The state militia system could be reformed, but it could not be replaced by a professional army. However, states reformed their militias in accordance with their "internal politics" and "need for an internal police." For example, the South reformed its militias in a way that served southern racism.[110] Officially, the South said it would revive its militias in order "to replace federal troops in the control of racial disturbances, lynchings, and vigilantism." In actuality, southern states did very little to strengthen their militias. In the North, on the other hand, the "primary function" of a revived militia was to serve as "a state police employed in the control of labor disturbances." Consequently, northern "states with large working-class populations took the lead in the militia revival."[111] Dominant forces within society influenced state government to reform militias in ways that would serve their interests. The civic militia no longer existed to protect the interests of "the people" against state-imposed tyranny.

Universal Military Service and the Attack on the Citizen-Soldier

With the successful building of the "new American state" during the first two decades of the twentieth century came a series of military reforms that ultimately gutted the normative ideal of the Citizen-Soldier by disconnecting military service from participatory citizenship. The first step in this process was the Dick Act of 1903, which repealed the Uniform Militia Act of 1792 and increased national power vis-à-vis the states' militias. Thinking in national rather than local terms, this act considered the military manpower potential of the "*whole . . . country*" and called it the "reserve militia," replacing the traditional term, "unorganized militia." (Henceforth the National Guard became the "organized militia.")[112] In addition, the Dick Act gave the federal government the power both to dictate how often National Guard units would drill and to inspect them. While National Guard units continued to be the military of first resort, before volunteers, once National Guard units were enlisted in federal service, they became part of the federal volunteer army and their integrity could not be guaranteed.[113] The Dick Act was followed in 1908 by a new militia bill that further expanded federal control over the National Guard units and established that they could be used outside of the United States.[114]

Finally, in the last years of the Progressive Era—and amidst much controversy—Congress passed two important acts that successfully finished the process of creating a *national* military that could serve the interests of the "new American state." First, the National Defense Act of 1916 consolidated and extended federal power over the National Guard. "The most comprehensive piece of military legislation ever passed by Congress," it enlarged the peacetime regular army and authorized the president "to institute a draft in wartime to fill out the ranks of the regular army." The National Guard maintained its position as the first-line offensive reserve, but it lost its rootedness in local communities, as it came under greater federal control. Guardsmen would now have to take an oath to serve *national* as well as state governments and would be expected to serve outside the United States if ordered.[115] In 1917 the Selective Draft Act completed this process by creating the Selective Service System to register all male citizens for a *national* draft.[116]

Despite its emphasis on the military obligations of all male citizens, the Selective Draft Act does *not* represent the epitome of the Citizen-Soldier tradition, because it does not link military service to participatory citizenship. That is to say, as I have argued throughout this study, the Citizen-Soldier constitutes a normative ideal that links military service to participatory citizenship; it is not merely an empirical description of a military made up of liberal citizens. Nor can it be reduced to the idea that citizens serve in the military only temporarily, as some have argued.[117] Rather, the Citizen-Soldier tradition makes soldiering central to the process of *becoming* a citizen, because martial practices instill in citizens the virtues *required* for participation in self-government aimed at the common good. Thus, because the Selective Draft Act did not connect the national draft with a reworked version of participatory citizenship suited for America's new status as a nation-state, it does not fully exemplify the political theory of the Citizen-Soldier.

Although the act does incorporate some elements of the Citizen-Soldier tradition, it must be understood as part and parcel of the watershed transition from the civic republican model of participatory citizenship to the liberal model of individualism that occurred over the course of the Progressive era.[118] That is, the Selective Draft Act drafted men into federal service as *individuals* rather than as members of a community-based militia. Thus, the act both came out of and fed into the increasing hegemony of liberal *individualism* in a newly unified society that had previously been characterized by civic republican ideals.[119] Though the American militia system did indeed need to change in light of twentieth-century contingencies, America's increasingly *liberal* milieu *undermined the possibility of reconfiguring civic republican ideals*—such as the linkage between military service and participatory citizenship—for a newly constructed national state. Military service would no longer go hand in hand with substantive participation in the process of self-government.

Nevertheless, the Citizen-Soldier ideal occupies such an important place in the American political tradition that it did not completely disappear. Despite the fact that the United States switched to an all-volunteer force in 1973, the specter of the Citizen-Soldier remains. Although the Citizen-Soldier ideal appeared to have died in the battles of the nineteenth century that led to liberalism's eclipse of the civic republican tradition, its ghost continues to haunt American political mythology even today, and—as we shall see in the next chapter—it seems to be transmuting into a sinister new form.

Notes

1. See Louis Hartz, *The Liberal Tradition in America* (San Diego: Harcourt Brace Jovanovich, 1955).

2. For the foundational works of the republican revision, see Bernard Bailyn, *The Ideological Origins of the American Revolution* (Cambridge: Belknap Press of Harvard University Press, 1992); J. G. A. Pocock, *The Machiavellian Moment: Florentine Political Thought and the Atlantic Republican Tradition* (Princeton: Princeton University Press, 1975); and Gordon S. Wood, *The Creation of the American Republic, 1776–1787* (Chapel Hill: University of North Carolina Press, 1969). For an important recent communitarian contribution to the republican school, see Michael J. Sandel, *Democracy's Discontent: America in Search of a Public Philosophy* (Cambridge: Belknap Press of Harvard University Press, 1996).

3. Benjamin R. Barber, "Unscrambling the Founding Fathers," *New York Times Book Review,* 13 January 1985, 9.

4. Barber, "Unscrambling the Founding Fathers," 10.

5. As Barber explains, liberal democracy is "linked in a single circle of reasoning that begins as it ends in the natural and negative liberty of men and women as atoms of self-interest, as persons whose every step into social relations, whose every foray into the world of Others, cries out for an apology, a legitimation, a justification." What all strands of liberalism share is "a belief in the fundamental inability of the human beast to live at close quarters with members of its own species." Consequently, liberalism seeks "to structure human relations by keeping men apart rather than by bringing them together. It is their mutual incompatibility that turns men into reluctant citizens and their aggressive solitude that makes them into wary neighbors." Benjamin R. Barber, *Strong Democracy: Participatory Politics for a New Age* (Berkeley: University of California Press, 1984), 20–21.

6. J. R. Pole, *The American Constitution For and Against: The Federalist and Anti-Federalist Papers* (New York: Hill and Wang, 1987), 14.

7. See Pole, *American Constitution,* 14.

8. Wood, *Creation of the American Republic,* 49–50.

9. Wood, *Creation of the American Republic,* 49.

10. Pole, *American Constitution,* 84.

11. Pole, *American Constitution,* 52.

12. Pole, *American Constitution,* 120.

13. John P. Kaminski and Richard Leffler, eds., *Federalists and Antifederalists: The Debate Over the Ratification of the Constitution* (Madison, WI: Madison House, 1989), 12.

14. Sandel, *Democracy's Discontent,* 28–39; Wood, *Creation of the American Republic,* 60.

15. Wood, *Creation of the American Republic,* 58.

16. Robert E. Shalhope, "The Armed Citizen in the Early Republic," *Law and Contemporary Problems* 49 (Fall 1986), 139. Shalhope convincingly argues that eighteenth-century Americans understood the right to bear arms as both an individual and a civic obligation.

17. For criticisms of the military effectiveness of the Militia Act of 1792 see Reuben Elmore Stivers, *Privates and Volunteers: The Men and Women of Our Reserve*

Naval Forces: 1766 to 1866 (Annapolis: Naval Press Institute, 1975), 173; and Theodore J. Crackel, *Mr. Jefferson's Army: Political and Social Reform of the Military Establishment, 1801–1809* (New York: New York University Press, 1987), 10. For a defense of the Act, see Russell F. Weigley, *History of the United States Army* (Bloomington: Indiana University Press, 1984), 94.

18. Weigley, *History of United States Army,* 93, emphasis mine.

19. Samuel P. Huntington, *The Soldier and the State: The Theory and Politics of Civil-Military Relations* (Cambridge: Belknap Press of Harvard University Press, 1957), 195.

20. Huntington, *The Soldier and the State,* 196.

21. Huntington, *The Soldier and the State,* 196–98.

22. Crackel argues that these positions are contradictory. Jefferson's "creation of a military school at West Point—the quintessence of that regular army he supposedly detested—cannot be made to fit this [Citizen-Soldier] mold." Crackel, *Mr. Jefferson's Army,* 1. Crackel argues that Jefferson created a military school in order to teach military officers to support republican ideals. Though this may be true, Jefferson did not rely simply on filling his army with individual supporters of republicanism. Instead, he helped establish a new version of the Citizen-Soldier tradition at the structural level. Because the "technicism" strand of American military theory disrupted the unity of the officer corps, it served as one of the key goals of the Citizen-Soldier tradition: It prevented the emergence of a professional military.

23. Political theorists often use the term *civil society* in very different ways. This book defines *civil society* in the "strong democratic" sense outlined by Barber—the space between state and market in which individuals engage together in civic practices. See Benjamin R. Barber, *A Place for Us: How to Make Society Civil and Democracy Strong* (New York: Hill and Wang, 1998).

24. Alexis de Tocqueville, *Democracy in America* (New York: New American Library, 1956), 61, emphasis mine.

25. Tocqueville, *Democracy in America,* 61.

26. Tocqueville, *Democracy in America,* 46.

27. Tocqueville, *Democracy in America,* 68–69.

28. Tocqueville, *Democracy in America,* 70.

29. Tocqueville, *Democracy in America,* 95.

30. Tocqueville, *Democracy in America,* 108.

31. Tocqueville, *Democracy in America,* 110.

32. Tocqueville, *Democracy in America,* 109.

33. Tocqueville, *Democracy in America,* 98.

34. Tocqueville, *Democracy in America,* 194.

35. Tocqueville, *Democracy in America,* 193.

36. Tocqueville, *Democracy in America,* 193.

37. Tocqueville, *Democracy in America,* 196.

38. Tocqueville, *Democracy in America,* 196.

39. Tocqueville, *Democracy in America,* 197.

40. Tocqueville, *Democracy in America,* 115.

41. Tocqueville, *Democracy in America,* 115.

42. Tocqueville, *Democracy in America,* 117.

43. Tocqueville, *Democracy in America,* 108.

44. Michael J. Sandel makes this point in *Democracy's Discontent,* 320.

45. The existence of the Citizen-Soldier tradition in America is well documented. For instance, see John Whiteclay Chambers II, *To Raise an Army: The Draft Comes to Modern America* (New York: Free Press, 1987); Lawrence Cress, *Citizens in Arms: The Army and the Militia in American Society to the War of 1812* (Chapel Hill: University of North Carolina Press, 1982); John K. Mahon, *History of the Militia and the National Guard* (New York: Macmillan, 1983); Allan R. Millett and Peter Maslowski, *For the Common Defense: A Military History of the United States of America* (New York: Free Press, 1984); and J. G. A. Pocock, "The Americanization of Virtue," in Pocock, *Machiavellian Moment.*

46. Catherine Bell, *Ritual Theory, Ritual Practice* (Oxford: Oxford University Press, 1992), 187.

47. Shalhope, "Armed Citizen," 139.

48. Crackel, *Mr. Jefferson's Army,* 5.

49. Shalhope, "Armed Citizen," 140.

50. Crackel, *Mr. Jefferson's Army,* 8; Chambers, *To Raise an Army,* 10.

51. Stivers, *Privates and Volunteers,* 173.

52. Stivers, *Privates and Volunteers,* 173. See also Millett and Maslowski, *For the Common Defense,* 5.

53. Chambers, *To Raise an Army,* 5–6.

54. Mahon, *History of the Militia,* 83–84; Chambers, *To Raise an Army,* 5–6.

55. Michael E. McGerr, *The Decline of Popular Politics: The American North, 1865–1928* (New York: Oxford University Press, 1986), 23, emphasis mine.

56. Mahon, *History of the Militia,* 83–85, emphasis mine.

57. McGerr, *Decline of Popular Politics,* 30.

58. McGerr, *Decline of Popular Politics,* 26.

59. McGerr, *Decline of Popular Politics,* 27.

60. McGerr, *Decline of Popular Politics,* 40.

61. McGerr, *Decline of Popular Politics,* 40.

62. McGerr, *Decline of Popular Politics,* 37–38, emphasis mine.

63. Lauren Berlant, *The Anatomy of National Fantasy: Hawthorne, Utopia, and Everyday Life* (Chicago: University of Chicago Press, 1991), 34, emphasis mine.

64. Michael Schudson, "Was There Ever a Public Sphere? If So, When? Reflections on the American Case," in *Habermas and the Public Sphere,* ed. Craig Calhoun (Cambridge: MIT Press, 1992), 145.

65. Schudson, "Was There Ever a Public Sphere?" 146.

66. Schudson, "Was There Ever a Public Sphere?" 159.

67. Richard R. Moser, *The New Winter Soldiers: GI and Veteran Dissent during the Vietnam Era* (New Brunswick, NJ: Rutgers University Press, 1996), 18.

68. See Chambers, *To Raise an Army,* 15; and Mahon, *History of the Militia,* chap. 2.

69. Mahon, *History of the Militia,* 86–87.

70. Mahon, *History of the Militia,* 93.

71. Mahon, *History of the Militia,* 94.

72. Mahon, *History of the Militia,* 94.

73. Chambers, *To Raise an Army,* 15.

74. Chambers, *To Raise an Army,* 37.

75. Chambers, *To Raise an Army,* 38.

76. See Chambers, *To Raise an Army,* 15; and Mahon, *History of the Militia,* chap. 2.

77. Allen W. Trelease, *White Terror: The Ku Klux Klan Conspiracy and Southern Reconstruction* (Baton Rouge, LA: Louisiana State University Press, 1971), xlii.

78. Trelease, *White Terror,* xlii. See also John Hope Franklin, *The Militant South* (Cambridge: Belknap Press of Harvard University Press, 1956), 72–73.

79. Chambers, *To Raise an Army,* 37.

80. David Osher, untitled paper presented at the American Historical Association annual convention (1993), 3.

81. Mahon, *History of the Militia,* 85.

82. Mahon, *History of the Militia,* 109; Trelease, *White Terror,* xliv.

83. Trelease, *White Terror,* xxxiv.

84. Mahon, *History of the Militia,* 109.

85. Trelease, *White Terror,* xxxv.

86. Mahon, *History of the Militia,* 109; Trelease, *White Terror,* xlv.

87. Mahon, *History of the Militia,* 109.

88. Trelease, *White Terror,* xlv.

89. Trelease, *White Terror,* xlv.

90. John Whiteclay Chambers II, *Tyranny of Change: America in the Progressive Era 1890–1920,* 2nd ed. (New York: St. Martin's Press, 1992), 1–6.

91. For a thorough discussion of these trends, see Chambers, *Tyranny of Change.*

92. Chambers, *To Raise an Army,* 38.

93. Stephen Skowronek, *Building a New American State: The Expansion of National Administrative Capacities, 1877–1920* (Cambridge: Cambridge University Press, 1982), 86.

94. Chambers, *To Raise an Army,* 38.

95. Skowronek, *Building a New American State,* 87.

96. Mahon, *History of the Militia,* 116.

97. "The demand for a military revival [between 1877 and 1900] was rooted in increasing class conflict and international capital expansion." Skowronek, *Building a New American State,* 87.

98. Skowronek, *Building a New American State,* 88.

99. For a fuller discussion of this transition, see R. Claire Snyder, "Shutting the Public Out of Politics: Civic Republicanism, Professional Politics, and the Eclipse of Civil Society," *An Occasional Paper of the Kettering Foundation* (Dayton, OH: The Charles F. Kettering Foundation, Spring 1999).

100. Skowronek, *Building a New American State,* 93, 91.

101. Skowronek, *Building a New American State,* 98.

102. Skowronek, *Building a New American State,* 91.

103. Skowronek, *Building a New American State,* 99.

104. Mahon, *History of the Militia,* 118.

105. Skowronek, *Building a New American State,* 100, emphasis mine.

106. Quoted in Weigley, *History of the United States Army,* 282.

107. Quoted in Weigley, *History of the United States Army,* 101–2, emphasis mine.

108. Mahon, *History of the Militia,* 116–18.

109. Skowronek, *Building a New American State,* 101–2, emphasis mine.

110. Skowronek, *Building a New American State,* 104.

111. Skowronek, *Building a New American State,* 105. For an oppositional view, see Weigley, *History of the United States Army,* 282.

112. Weigley, *History of the United States Army,* 321.

113. Skowronek, *Building a New American State,* 208; and Mahon, *History of the Militia,* 139. For a full discussion of the Dick Act, see Weigley, *History of the United States Army,* 320–24.

114. Skowronek, *Building a New American State,* 218; Mahon, *History of the Militia,* 139; Weigley, *History of the United States Army,* 324.

115. Skowronek, *Building a New American State,* 232.

116. For a much more detailed discussion of the political debates over military policy that led to the passage of the Selective Draft Act, see Chambers, *To Raise an Army.*

117. "From the original obligation to train and serve briefly in the colonial or state militia for *local* defense when necessary, Americans eventually derived the concept of temporary *national* wartime armies composed largely of amateur citizen-soldiers. . . . What the ideal of the Citizen-Soldier meant to Americans in practice was that every young adult male citizen or declarant alien might be *liable* to the military obligation for *temporary* service in what was agreed by public authorities to be an emergency. Initially and briefly, this service was in the colonial militia; by the 20th century, it was in a national wartime army." Chambers, *To Raise an Army,* 266.

118. For a discussion of this transition, see Sara M. Evans and Harry C. Boyte, *Free Spaces: The Sources of Democratic Change in America* (Chicago: University of Chicago Press, 1992), and Snyder, "Shutting the Public Out of Politics."

119. Mahon, *History of the Militia,* 155–56.

Chapter 5

Citizen-Soldiers, Blood Brothers, and the New Militias: Interrogating the Republican Discourse of the American Right

A new social movement is developing in America out of the quasi-fascist fringes of the American Right. Sparked by the federal government's attack on the Weaver family in Ruby Ridge, Idaho, in 1992 and its assault on the enclave of the Branch Davidian religious sect in Waco, Texas, in 1993, and fueled by the passage of the Brady Bill in 1993, the New Militias are organizing to defend themselves against what they see as the increasing tyranny of the American federal government. These New Militias envision a future of war, martial law, and one-world dictatorship.

The strange conspiracy theories advanced by some of these groups received widespread media coverage in the wake of the 1995 bombing of the federal building in Oklahoma City. What puzzles me most about their rhetoric is the use of the traditionally democratic discourse of the Citizen-Soldier to bolster their antidemocratic claims. Not only are the New Militias focusing increasingly on the individual's right to bear arms, but they are also reviving traditional civic republican concerns about the dangers of a standing army—antifederalist arguments that emphasized the importance of a civic militia in defending republican liberty.

Despite their excesses, these groups do raise some legitimate issues. For example, we might not in fact want the government to have a monopoly on the bearing of arms. So it is not their insistence on the importance of the Second Amendment that puzzles me. Instead it is the use of a traditionally democratic discourse—the discourse of the civic militia—for the purpose of advancing the neo-Nazi and white supremacist agenda that underlies much of the New Militia movement. While not every supporter of the New Militias supports racism and anti-Semitism, key organizers of the New Militia movement come directly out of the quasi-fascist fringe groups on the far Right of American politics, and their antidemocratic agenda still underlies what now masquerades as a

107

democratic movement of citizen-soldiers. It is my contention that America's racist and anti-Semitic right-wing groups have been able to build support for their agenda by "masking" their claims with two democratic discourses characteristic of the American political tradition—"identity politics" and the Citizen-Soldier tradition.[1] In so doing, they have rendered their ideas more palatable to those who do not want to be known as quasi-fascists, in particular, the "angry white men" who feel scared, economically and politically threatened, and shut out of the American political process.

A Crisis of Legitimacy

The New Militia movement must be understood within the broader context of American politics, and particularly in relation to the "culture wars" and the "crisis of legitimacy." In the first place, the "culture wars"—the fundamental disagreements between Christian conservatives and secular progressives about how to evaluate the changes inaugurated by the democratic social movements that came out of the 1960s—currently constitute a fundamental dynamic of American politics.[2] Many on the Right oppose the legal, political, and social changes that have resulted from the civil rights movement, the sexual liberation movement, the feminist movement, the youth movement, and the gay/lesbian rights movement, all of which profoundly challenge traditional societal hierarchies. Additionally, many conservatives oppose the progressive policies that have been instituted in order to create some sense of dignity and social justice for all the diverse groups that comprise multicultural America. In short, the conservative side in the "culture wars" has created a reactionary politics of backlash.

As a result of these fundamental conflicts, people can no longer imagine that a consensus over basic values underlies American society, a situation that often generates distrust and anxiety. In fact, as Amy Gutmann and Dennis Thompson have noted, "of the challenges that American democracy faces today, none is more formidable than the problem of moral disagreement. Neither the theory nor the practice of democratic politics has so far found an adequate way to cope with conflicts about fundamental values."[3] Various groups of citizens often fear and distrust one another, and in times of real or imagined economic insecurity, tensions worsen.

The conflicts produced by the "persistence of moral disagreement" are exacerbated by the increasingly widespread "crisis of legitimacy" that also permeates American society. That is to say, within a democratic context, legitimate government requires the consent of the governed. Yet in reality the government no longer has the full consent of the governed, because citizens feel that the government is both out of touch with them and in the service of special interests. In a democracy, citizens create government to serve the common good; they authorize the government to act on their behalf. Yet citizens across the political spectrum feel no sense of control over government and believe that government no longer serves the common good.[4] Moreover, because we no longer have an active tradition of participatory citizenship in this country, as I argued in chapter 4, large numbers of American citizens—progressives as well as conservatives—feel alienated, angry, and powerless over the changes and conflicts that our society is undergoing—changes and conflicts inherent in the very process of democracy itself. The suspicion that exists among different sectors of society simply exacerbates the "crisis of legitimacy." Citizens believe that government has been captured by "special interests," but who constitutes those "special interests" differs: Are they large corporations? Racist white people? Christian fundamentalists? Secular humanists? Radical homosexuals? Unqualified minorities? Or Jews? It depends who you ask.

This chapter argues that this broadscale crisis of legitimacy combined with the reactionary opposition to the progressive democratic changes achieved by the "new social movements" of the 1960s and their subsequent politics of identity have created an opportunity for quasi-fascist fringe groups on the edge of American politics—the neo-Nazis who have always been marginalized and the Ku Klux Klan that used to be a central political player—to build support for their agenda among more mainstream conservatives. These groups have been able to "mask" their racist and anti-Semitic agenda behind two democratic discourses that are central to the American political tradition: the currently dominant "identity politics" and the historically important Citizen-Soldier tradition. The New Militia movement represents their successful attempt to build a coalition among various groups of angry white men and their supporters.

Quasi-Fascist Roots

The quasi-fascist fringe groups that gave birth to the New Militia movement include neo-Nazi organizations, such as Aryan Nations, the National Alliance, and the Order; white supremacist groups like the Ku Klux Klan; and a variety of Aryan Christian organizations, such as Christian Identity and the survivalist group, the Covenant, the Sword, and the Arm of the Lord.[5] Despite their differences, all these groups share a white supremacist, anti-Semitic, Aryan Christian worldview, and all stand opposed to the liberal democracy that brought us a strong federal government, a social welfare state, the civil rights movement, the feminist movement, and the ongoing struggle for lesbian/gay civil rights. As opposed to identity groups devoted to securing the rights of previously and currently excluded groups, the New Militia movement's progenitors are antidemocratic advocates of racism and violence.[6]

The quasi-fascist Right continues a long tradition of antidemocratic resistance to multiculturalism that has plagued American society ever since it started to become truly multicultural at the end of the nineteenth century.[7] Like their

Nativist and xenophobic forebears, today's quasi-fascist groups fear and oppose the changes and conflicts inherent in the transition to an increasingly multicultural democracy. Relying on standard Nazi rhetoric, today's National Alliance makes *Jews* the symbol of multiculturalism and of what they see as the
excesses of democracy:

> Jews come into any homogeneous society—and such was America at the begin-
> ning of this century—as outsiders, as strangers. . . . To make way for themselves,
> to open up possibilities for penetration and control, they must break down the
> structure of the society, corrupt its institutions, undermine its solidarity, weaken
> its sense of identity, obliterate its traditions, *destroy its homogeneity.* Thus they in-
> evitably will be *in favor of democracy,* of permissiveness, of every form of self-
> indulgence and indiscipline. They will be proponents of *cosmopolitanism,* of *egal-
> itarianism,* of *multiculturalism.* They will oppose patriotism (except when they are
> inciting their hosts to fight a war on behalf of Jewish interests). They will agitate
> endlessly for *change, change, change,* and they will call it *progress.*[8]

However, what the National Alliance sees as the excesses of democracy are
simply its necessary corollaries—egalitarianism, inclusivity, diversity, multiculturalism, and "change, change, change."

In another passage, the National Alliance uses the standard Nazi tale about
a "Jewish conspiracy," first to explain the changes that American society has
undergone since it actually began to apply its democratic principles to all its citizens, and then to attack the federal government, which is responsible for instituting those changes:

> With the growth of mass democracy (*the abolition of poll taxes and other qualifi-
> cations for voters, the enfranchisement of women and of non-Whites*), the rise in
> the influence of the mass media on public opinion, and the insinuation of the Jews
> into a position of control over the media, *the U.S. government was gradually trans-
> formed into the malignant monster it is today:* the single most dangerous and de-
> structive enemy our race has ever known. Many patriots look back fondly at the
> *government as it was in its first phase,* when it was less democratic and *less intru-
> sive in the lives of citizens.*[9]

Because of the Jews, they tell us, the American people

> are presented with a single view of the world—a world in which *every voice* pro-
> claims the equality of the *races,* the inerrant nature of the *Jewish "Holocaust" tale,*
> the wickedness of attempting to halt a *flood of non-White aliens* from pouring
> across our borders, the danger of permitting citizens to keep and bear arms, the
> moral equivalence of all *sexual orientations,* and the desirability of a *"pluralistic,"*
> cosmopolitan society rather than a homogeneous one.[10]

Aryan groups reject multiculturalism, religious pluralism, and the progressive
extension of democratic principles to previously excluded groups, including

African Americans, Latinos, women, gay men, and lesbians. Although clearly extremists, these neo-Nazi groups speak to the same anxieties that underlie more moderate political movements, such as the "culture wars" and the movement for government devolution.

Aryan groups see the right to bear arms as central to their fight against democracy, and they try to build support for their agenda by criticizing institutions that many Americans now distrust or consider illegitimate, like the mass media, the United Nations, and the public schools, and public policies that are controversial, like sensitivity training and gun control:

> While posing as a public-spirited "civil rights" group, [the Anti-Defamation League of B'nai B'rith has] been working for decades to disarm law-abiding Americans, to *control our sources of news* and other information, and enslave us under a totalitarian *world government* which many have come to call the "New World Order." They do this through overt and covert propaganda, the creation of humanitarian-sounding front groups secretly controlled by the ADL, by the conducting of brainwashing sessions called *"sensitivity-training"* for members of our police forces, by the production and *introduction into the public schools of ADL propaganda* as "textbooks" or "resource material for teachers," and by their cozy relationship with *the controlled media,* which routinely print and broadcast ADL propaganda releases as so-called "news."[11]

Despite the pall of conspiracy, this passage illustrates the ways in which Aryan groups appeal directly to the concerns of the many mainstream American citizens who distrust institutions like the mass media, the United Nations, and the public school system, and who oppose liberal policies like gun control and sensitivity training.

Right-wing fringe groups also try to "mobilize and amplify [the] widespread anxieties, traumatic experiences, and unfulfilled wishes" of economically threatened white men who can no longer depend on white privilege to secure what they consider to be their rightful place in society.[12] For example, one piece of National Alliance propaganda features a picture of a beautiful blond-haired, blue-eyed little boy underneath which is the following caption: "MISSING: A Future for White Children in America. Description: blonde, brown, or red hair, clear eyes, intelligent, inquisitive, healthy, playful. *Future abducted by hateful minorities and corrupt politicians."*[13] Here Aryan groups fuel the anxieties of disadvantaged whites, appeal to conservative opponents to affirmative action, and attempt to exploit the widespread belief that our supposedly democratic government has been captured by corrupt politicians who pander to special interests.

As a solution to what they see as the serious problems of a multicultural democracy, quasi-fascist fringe groups want to turn the clock back to a time when only white, Anglo-Saxon Protestants had power: they want to create a movement for white ethnic nationalism. For example, the leader of Aryan Nations, Richard Butler, moved to Idaho to be with white people like himself. "Race" to Butler meant "nation," and he told his followers that no race of people could survive without a territory of its own. Other races and ethnic groups had nations in which to propagate. . . . He aspired to a nation-state for whites. . . . In time, Butler formulated the goal of Aryan Nations: to establish a state representing the voice and will of the white Aryan race as a divinely ordained, sovereign, independent people, separate from all alien, mongrel people in every sphere of their individual and national life.[14]

The National Alliance embraces a similar plan of action:

An essential element of the National Alliance message is . . . : We White Americans can pull ourselves up from the gutter of moral depravity and "multiculturalism" only after we have regained control of our news and entertainment media. . . . We must build a new government, based on racial principles and answerable only to White Americans, which provides real leadership instead of pandering to every racial minority, organized perversion, and special interest.[15]

Going well beyond the concerns of many, more moderate conservatives, the quasi-fascist Right opposes every principle of democracy—at least when extended to anyone other than themselves.

Fortunately for proponents of democracy, Aryan arguments include key elements that stand in direct opposition to American political traditions, and this limits the appeal of their arguments. First, the Aryan Right opposes both individualism and humanism, because they fragment efforts on behalf of the race. For example, the National Alliance specifically states that their philosophy "is in contrast to the attitude of the individualists, who do not recognize a responsibility to anyone but themselves; and to that of the humanists, who eschew their racial responsibility."[16] This critique runs directly counter not only to America's strongest philosophical tradition of liberal individualism but also to its main philosophical rival, the humanist discourse of civic republicanism.

Second, the Aryan Right favors a *citizenship of blood* (*ius sanguinis*) over the liberal *ius solis* and the civic republican *citizenship of civic practices*.[17] As discussed previously, *ius solis* defines citizenship by residence within a particular bounded territory, whereas a *citizenship of civic practices*—the focus of this study—understands citizenship as constituted through engagement in civic practices. In opposition to both of these democratic forms of citizenship, a *citizenship of blood* restricts citizenship to people who have a particular familial, ethnic, or racial heritage—a particular type of blood. The Aryan Right clearly espouses the latter: "If the White race is to survive we must unite our people on the basis of common blood."[18] Again, this position puts Aryan groups at odds with both liberalism, with its *ius solis,* and civic republicanism, with its *citizenship of civic practices.*

Finally, the Aryan solution to what they consider to be the evils of democracy requires the creation of a strong state, an idea that not only runs counter to contemporai y public opinion but also stands in tension with the American political tradition of limited government. As the National Alliance explains,

> Perhaps the time will come when we can afford to have a minimal government once again, but that time lies in the remote future. The fact is that we need a strong, centralized government spanning several continents to coordinate many important tasks during the first few decades of a White world: the racial cleansing of the land, the rooting out of racially destructive institutions, and the reorganization of society on a new basis.[19]

Quasi-fascist arguments like this one that clearly contradict well-entrenched political beliefs have limited appeal. Because this type of far Right rhetoric does not appeal to mainstream Americans who are individualistic and suspicious of a powerful national state, building a movement requires repackaging this agenda.

"Red Neck, White Skin, and Blue Collar": Identity Politics for Angry White Men

The New Militia movement arose out of the attempt of quasi-fascist fringe groups to "mask" their agenda behind two characteristically American political discourses that have more populist appeal: "identity politics" and the Citizen-Soldier tradition. In the first case, the far Right has been quite successful in repackaging its racist claims as an "identity politics" for angry white men. That is, many white men have come to see themselves as "targets of all other empowerment movements, from women's liberation to black power to gay pride"—targets of "identity politics"—and so are receptive to the idea that they need their own version of identity politics.[20] For example, Bob Mathews, founder and leader of the Order, an offshoot of Aryan Nations, "gradually . . . accepted racism not as a doctrine of hate for races but as a matter of pride and love for his white race. Blacks, browns, and others celebrated their racial identity, so it couldn't be wrong, he considered, for whites to feel good about their race as well."[21] Mathews wanted "to radicalize" the "latent racism in a great mass of the white middle class."[22] He wanted to build a movement out of various groups of whites who "distrust the government," including "Klansmen, neo-Nazis, survivalists, tax protesters, militant farmers, Identity churches, and other groups." They would be "the new 'red, white, and blue'—red neck, white skin, and blue collar."[23] By framing this argument in terms of "identity politics," the racist Right legitimizes its agenda so that even those not interested in joining the movement—including those who stand opposed to the movement—nonetheless recognize the logic of its claims.

Still, it is important to note that this far Right version of identity politics is *not* a democratic movement but rather a *backlash reaction against democracy.*

That is to say, groups advocating this new "identity politics" do not portray white men as simply another particular identity among the many that make up multicultural America. Instead, they want to reclaim the privileges and entitlements that white men enjoyed prior to the (ostensibly) full enfranchisement of women and minorities. For example, participants in an Aryan study group on the World Wide Web begin by identifying themselves as "European Americans."[24] But instead of simply exploring the cultural traditions of their particular people, the group quickly notes that they are "angry at the United States Government and at the costs and disadvantages it imposes on European Americans. The legal and economic burdens fall predominantly on the young and politically weak; those seeking college admission, first time job seekers, and entrepreneurs starting up businesses." These "European Americans" see themselves not as one important part of a multicultural society but rather as its victims. Because of "racial quotas," they argue, white European American males are being forced out of the public sector. As a consequence the "public sector bureaucracies will become overwhelmingly black, brown and female over the next 20 years. . . . Every contact with government will become contact with non-Europeans over the next 20 years." These European Americans oppose attempts to make our supposedly democratic government mirror the people whom it represents.[25]

Building a Movement: Citizen-Soldiers on the Right

The Citizen-Soldier tradition forms the second important characteristically American discourse used by the quasi-fascist Right to "mask" its agenda. While "identity politics" politicizes the particular location of white males within society, the Citizen-Soldier tradition helps form these disparate individuals into a coherent movement. In essence, the New Militia movement mobilizes to advance the interests of this newly politicized identity group, Angry White Men. And that is why, for the most part, economically threatened white men comprise the New Militias.[26]

Although the New Militia movement began in earnest in 1993 with a call for a return to "the Unorganized Militia of the United States of America,"[27] which traditionally included all able-bodied American males, and grew so quickly that by "the spring of 1994 the new militias appeared spontaneously and simultaneously from coast to coast,"[28] the first militia groups actually formed during the 1960s, in direct reaction to the democratic changes that were afoot. For example, a secret group called the Minutemen organized during the 1960s "to promote their position that whites should arm themselves, practice urban combat and otherwise prepare for a dismal future that promised extensive race riots at best and all-out thermonuclear war at worst."[29] Using the rhetoric of the Citizen-Soldier tradition, these staunch anti-Communist racists armed themselves to protect what they saw as *their* Republic from internal and external enemies—both real and imagined. The Minutemen called for "an extensive investigation" into

> the loyalty of all the officials in government, defense industry, tax free foundations, labor unions, the communications industry, news media and similar fields vital to the nation's internal security, such investigations being made for the most part by Grand Juries composed of private citizens of substance and good repute. . . . These investigations should include not only the Departments of State, Defense, Health, Education, and Welfare, Agriculture and Labor but also the Treasury, the Justice Department, the F.B.I. and the C.I.A.[30]

Part of the antidemocratic reaction to the democratic changes of 1960s, the Minutemen stood opposed to the power of working people, to the passage of civil rights legislation, to the progressive reforms of the Great Society, and to the federal government that advanced these causes.[31] Yet they mobilized using the democratic rhetoric of the Citizen-Soldier tradition.

Like the New Militia movement today, the Minutemen believed that the American government had become illegitimate, and they were especially angry with the Supreme Court, which ended legal segregation and mandated racial integration. "Where evidence of treason is found," they argued,

> those suspected of such acts should be tried before civilian courts but if it is found that *the courts themselves have been infiltrated to such an extent as to make the conviction of traitors impossible,* then the Constitution should be amended to allow such persons to be tried before military courts or before new federal judges especially appointed for that purpose. . . . [Moreover,] *all the present members of the Supreme Court should be removed from office.*

Moreover, like militia members today, these self-proclaimed citizen-soldiers condemned America's participation in the United Nations and wanted to strengthen "our nation's Christian heritage."[32] While the Minutemen advocated a very activist version of popular sovereignty, they tied this ideal to the creation of some sort of Christian nationalism. Indeed, the Minutemen were anti-Semitic and white supremacist as well as anti-Communist.[33] Clearly, the Minutemen stand as the direct forebears of the most antidemocratic, racist elements of the New Militia movement today.

Another early group that used the democratic rhetoric of the Citizen-Soldier ideal to advance a reactionary agenda was the Posse Comitatus, founded in 1969. Interestingly, the name *posse comitatus* comes from the German barbarian tribes who occupied Western Europe after the fall of the Roman Empire and who laid the foundations for feudalism. Although *posse comitatus* means literally "the power of the community," it does not represent a *democratic* version of this idea. Instead, the German barbarian tradition of *posse comitatus* "was grounded in ceremonies of fealty which through vows and ceremonies *subordinated* the individual to the tribe. The key tenet was the obligation of each member of the tribe to make himself a part of the tribe's power; specifically, each member was at all times to answer his chieftain's call to arms."[34] Although the tradition of *posse comitatus* came to America via English feudal law, the tradition is *distinctly* different from the Citizen-Soldier tradition of civic republicanism that also came to America via England, but which originated in classical humanist ideals that were modernized by Machiavelli. Thus, even the name of the group, Posse Comitatus, misinterprets the nature of the Citizen-Soldier tradition, which is democratic rather than authoritarian.

Long before it was popular, Posse Comitatus argued that American institutions had become illegitimate. Attempting to ground its arguments in American political theory, the Posse proclaimed that

> since the formulation of our Republic, the local County has always been the seat of government for the people. A county government is the highest authority of government in our Republic as it is closest to *the people, who are in fact, the government.* The county Sheriff is the only legal law enforcement officer in the United States of America. He is elected by the people and is directly responsible for law enforcement in his County. *It is his responsibility to protect the people of his County from unlawful acts on the part of anyone, including officials of government.* . . . The Sheriff is accountable and responsible only to the citizens who are the inhabitants of his County.[35]

The sheriff must protect all citizens, and "this protection extends to Citizens who are being subjected to unlawful acts even by officials of government, whether these be judges or courts or Federal or State agents of any kind whatsoever."[36] Again we see the belief that our democratic government has become illegitimate and tyrannical.

The Posse Comitatus justifies itself in terms of the Second Amendment and claims to be the traditional unorganized militia: "The Posse and the Militia have essentially the same purpose; they are men who act in the execution of the law."[37] That is to say,

> the Sheriff can mobilize all men between the ages of 18 and 45 who are in good health and not in the federal military service. OTHERS CAN VOLUNTEER! This body of citizens is the Sheriff's Posse. Each must serve when called by the Sheriff. The title of this body is the Posse Comitatus. The Posse is the entire body of those inhabitants who may be summoned by the Sheriff, or who may volunteer, to preserve the public peace or execute any lawful precept that is opposed. Since the Sheriff is the servant of the citizens who are inhabitants of the County, it is not his choice as to whether or not the Posse is organized and brought into being. It is only his choice as to whether or not he wishes to use it.[38]

But while the Posse, like the Minutemen, uses the discourse of democracy, it mobilizes in direct reaction to the extension of democratic principles to African

American citizens. In other words, the democratic discourse of the Citizen-Soldier ideal has been used by those actually opposed to democracy as a way of "masking" their agenda for many years. However, only recently has it begun to resonate with increasing numbers of mainstream Americans.

Interestingly, although a part of the racist, anti-Communist Right of the 1960s, the Posse Comitatus today actually claims to speak for working people. Whereas in the 1960s the Posse Comitatus, as well as Minutemen, was concerned primarily with resisting racial integration, which was being imposed by the federal government and the courts, today the Posse's concerns have expanded to issues of concern to working people threatened by environmental regulations.[39] Thus, the Posse now "insists that county sheriffs have the right to arrest federal land managers who fail to respect the 'customs and culture' of logging, mining and grazing on public lands."[40] The old Posse Comitatus has evolved into the growing "Counties Movement," a central component of the New Militia movement. The basic claim remains the same—they are not subject to any authority higher than the county sheriff—and their basic purpose also remains unchanged—to protect the interests of angry white men who view the institutionalization of democratic reforms that threaten their way of life as a form of tyranny.

The discourse of the Citizen-Soldier is particularly useful for the far Right, because it fulfills two key purposes. First, the Citizen-Soldier ideal conveniently provides the reactionary Right with a constitutionally protected way to arm its followers. Second, the discourse of the Citizen-Soldier has proven to be an effective way of bridging the gap between those on the far Right and the center Right. That is to say, the organizers of the New Militia movement have repackaged much of their reactionary agenda, "masking" it with the rhetoric of the Citizen-Soldier tradition, in order to render their ideas more palatable to economically threatened and politically alienated white men who are less extremist. As one victim of neo-Nazi violence puts it, the New Militia movement is "using oppressive big government as the come-on. . . . It's a new hook, a new way to recruit people and get attention."[41] In fact, watchers of the Right in America "all generally agree" that the explicitly racist and anti-Semitic programs of the "early militias acted as automatic checks on their growth. . . . [B]ut in the past few years, against a general background of economic uncertainty and alienation from the political system, the militias have made a radical turn *toward* the mainstream."[42] But while more moderate Americans are joining the New Militia movement, so are extreme racists. According to Joe Roy of Klanwatch, "the militia phenomenon that's sweeping the country. . . . [is] soaking up a lot of potential Klan members."[43] Thus, the New Militia movement appeals to a broad spectrum of angry white men, from hard-line quasi-fascists to white supremacists to conservative Republicans to those who are politically and economically marginal.

Not only have quasi-fascist groups been able to "mask" their extremism with the rhetoric of the Citizen-Soldier tradition, but these groups use other kinds of

code language as well. For instance, in making frequent reference to a conspiracy of "international bankers," the New Militia movement clearly invokes
the traditional anti-Semitic belief in a conspiracy of Jewish bankers.[44] In other
words, they do not actually have to say the word "Jew" to conjure up the same
feelings of paranoia that have historically fueled anti-Semitism. By making
"clearly recognizable" references without the explicitly anti-Semitic language,
the Militia movement can "escape blame."[45] Similarly, many advocates of the
"common-law courts" movement deny the legitimacy of any amendment added
after the Bill of Rights, without acknowledging the direct connection of this denial to Christian Identity theology—a movement that views the Declaration of
Independence, the U.S. Constitution, and the Bill of Rights as divinely revealed
Truth and every amendment after the Tenth as a Jewish conspiracy to destroy
the white race.[46] In the words of one observer, "instead of openly caviling
against Jews, blacks and other targets, Populist literature employs code words
linked to powerful themes that also can be heard in the American political
mainstream, such as questioning the wisdom of spending billions on foreign aid
(especially the aid to Israel) when America's own farmers are going bankrupt
in record numbers because they can't repay federal loans."[47] As another observer puts it, "most of the conspiracy theories that whirl about the militia
movement see the same evil cabal of bankers, politicians, and the media—but
without reference to ZOG [the Zionist Occupation Government] or Jewish control. *Many in the militia movement who believe these theories appear to be un-
aware that they were first put forth by white supremacists.*"[48] In other words,
quasi-fascists have been able to "mask" their anti-Semitic ideas behind more
moderate-sounding code language and thus build support among those who do
not want to be known as anti-Semites.

In fact, quasi-fascist groups have been so successful at repackaging their
ideas for mainstream political debates that the New Militia movement's political platform now simply looks like a more extreme version of the Republican
Party platform. For example, neo-Nazi groups traditionally advocate a *citizenship of blood* and so demand that we should "make citizenship a proud privilege to be earned, not a right carelessly awarded simply by birth in a certain ge-
ographical area."[49] This demand directly violates the Constitution, runs counter
to the liberal American view of citizenship (*ius solis*), and used to be considered an extremist position, outside of mainstream political debate. But now the
mainstream Republican party makes a similar argument. For instance, at their
1996 national convention, the GOP decided that the country should rescind the
part of the Fourteenth Amendment that says, "all persons born or naturalized
in the United States and subject to the jurisdiction thereof, are citizens of the
United States and of the State wherein they reside" (Section 1), in order to prevent the children of illegal aliens from reaping the benefits of American society. This demand chips away at the *ius solis* and lays the groundwork for a *citizenship of blood* (*ius sanguinis*). Furthermore, David Duke's quasi-fascist

"Populist Party" platform from 1984 looks remarkably similar to much of the current Republican Party platform.[50] Or, to be more specific, Duke's platform is the version of the Republican platform Pat Buchanan ran on in the 1996 primaries and from which Bob Dole (sometimes) tried to distance himself.

In other words, while the New Militia movement has repackaged quasi-fascist demands for mainstream consumption, it has also radicalized mainstream Republican beliefs. For example, many moderate Republicans (and others) believe that states should have more power vis-à-vis a federal government that is bloated and out of control. Pushing this belief further Right, the New Militia movement proclaims that the federal government has become so large that it has overstepped its constitutional authority.[51] And in even more extreme terms, the New Militia movement goes so far as to argue that the U.S. federal government is a tyranny conspiring with various other entities to take away the traditional freedoms of the American people. While the milder version of this claim—that states should have more rights—has resulted in attacks on large sectors of the federal government, the harsher version has fueled not only the "counties movement" but also the "common-law courts" movement which "rejects state and Federal statutes and all constitutional amendments except the Bill of Rights"—once again advocating a major tenet of Christian Identity. In fact, this latter movement, now operating in 40 states—and supported by such people as the infamous Montana Freemen—constitutes "the fastest growing sector of the far right Patriot movement," according to "law-enforcement officials and human rights organizations."[52] Far Right views obviously resonate with increasing numbers of American citizens.

The beauty of the New Militia movement is that it has been able to form a coalition among various sectors of the American Right.[53] White supremacists and neo-Nazis now join forces with other ostensibly less reactionary right-wing white men who are concerned about a broad range of conservative Republican party concerns, including opposition to the federal government, environmentalism, illegal immigration, taxes, foreign aid, welfare, affirmative action, crime, abortion choice, secular humanism, feminism, lesbian/gay rights, and, of course, gun control. According to an article in *The Nation,* observers of the American Right "all generally agree" that "the militias are a movement that for some time has been waiting to happen, born in the backlashes against civil rights, environmentalism, gay rights, the pro-choice movement, and gun control."[54] Thus, the New Militia movement actually brings together many diverse groups of angry white men and their supporters.

For all its faults, the New Militia movement does speak to some Americans who have legitimate political and economic grievances. The economic woes of New Militia supporters are very real. New Militia supporters include many working-class men and farmers, two constituencies that were seriously hurt by the tax cuts and military spending of the Reagan administration.[55] As another author puts it, "smack-dab in the middle of the American continent is a group of people who really are being driven into poverty by a system under which outside bankers foreclose on their government loans, force them to sell their hard-earned property at public auction, then drive them off their land jobless, penniless and unwanted by their equally beleaguered neighbors."[56] The New Militia movement has been very successful at capitalizing on the economic hardships of angry white men.

Among bankrupt farmers and displaced workers, the racist Right, which had been in decline since the end of the large-scale struggle against integration led by George Wallace and his supporters, was able to find many supporters for its anti-Semitic explanation for market readjustments and democratic reforms.[57] "The Anti-Defamation League of B'nai B'rith views the Populist Party's campaign to exploit the farm crisis as the most sophisticated political move by the anti-Semitic right in recent memory."[58] Unfortunately, the American Left has not done a very good job of speaking to those brought into the rank and file of the New Militia movement.

Nevertheless, as the author of one study of the New Militia movement writes: "It is important to note that not all militia members believe the far-out conspiracy theories. . . . Many militia members represent . . . mainstream discontent . . . , and others are active because of a more narrowly defined concern about gun control. It's easy to dismiss the paranoid extremists, but the discontent driving the militia movement is real and extends far beyond the membership of citizens' militias."[59] Supporters of the New Militia movement, like supporters of European fascism at the beginning of the twentieth century, are not all hard-core ideologues. European fascism spread throughout the populous under conditions of "widespread unemployment, an impoverished middle class, a terrorized petite bourgeoisie"—the very conditions that mobilize militia members in the United States today.[60]

But while not all members of the New Militias support racism and anti-Semitism, it is also important to acknowledge that the movement grew directly out of the quasi-fascist Right and often simply "masks" that extreme agenda with ostensibly democratic rhetoric. For example, the Militia of Montana (M.O.M.)— also known as the "Mother of All Militias"—which played a key role in stimulating the emergence of militias throughout the country, was founded by John Trochmann, a well-known advocate of the racist, anti-Semitic Christian Identity movement.[61] In order to expand his base of support, Trochmann and other militia members have

"cooled it on the religious fervor side and upped the heat on the political diatribe side." Trochmann had been trying to get a political movement going ever since his onetime friend Randy Weaver had his deadly showdown with federal authorities in August 1992. With the militia movement he found a way to attract a wider audience for his conspiracy theories; he just needed to, in his words, "leave religion at the door."[62]

Although many of the militias try to distinguish themselves from hate groups, such as the Klan, as *Time* magazine concluded, "such distinctions . . . are not always apparent."[63] The *New York Times* cites a federal investigator on this point: "All these groups, if you put them in a bag and shook them out, you couldn't tell one from the other."[64] Again, the New Militia movement constitutes a repackaging of quasi-fascist ideals for mainstream American consumption.

Even the more reasonable sounding Michigan militia, in which members took an oath to "defend the Constitution of the United States against all enemies, both foreign and domestic" and which sees itself as "a public, uniformed brigade of citizen soldiers," espouses thinly masked versions of typical quasi-fascist conspiracy theories.[65] For example, "Mark from Michigan," who "was one of the first to aggressively encourage the formation of citizens' militias" spins the following tale of conspiracy and tyranny:

> Elements within the U.S. government are working with foreign leaders to turn the United States into a dictatorship under the leadership of the United Nations. The battle to create this U.N. dictatorship, known as the new world order, is already well under way. Foreign troops are already training on American soil for a planned attack on Americans who resist; a network of forty-three detention centers has been set up throughout the country to serve as concentration camps for the resisters; plans have been readied to control the population through microchips to be implanted in newborn babies and through radio boxes already in place in automobiles made after 1985.[66]

This story sounds very similar to the one presented in the favorite quasi-fascist novel of the new militiamen, *The Turner Diaries*—the novel that inspired Tim McVeigh to bomb the federal building in Oklahoma City.[67] Although Mark from Michigan's narrative elides all explicit references to Jews and African Americans, it would obviously appeal to the same sorts of people as would quasi-fascist tales.

Let's compare Mark's narrative to the following one: "The U.S. government has been taken over by a conspiracy of Jewish bankers and nebulous other dark forces who plan to bleed the country dry, then bring a nuclear attack down upon the withered shell."[68] In the meantime, the Jews, in conjunction with the United Nations, are engaging in "mind-control tactics to force white American men to work at slavish menial jobs while making sure, through the income tax and affirmative action laws, that they [cannot] earn enough money to better themselves, even as the government [is] helping African Americans get better jobs and more training."[69] And of course, the federal government is attempting to disarm American "patriots" so that they cannot defend themselves against this impending tyranny. Many sundry versions of such conspiracy theories were exposed by the media in the wake of the Oklahoma City bombing.

While both of these conspiracy tales seem patently absurd, compare them to the National Alliance's more factually grounded interpretation of American politics:

For many years the power elite have pursued a purposeful policy of turning America into a nonwhite nation. By *changing our immigration* laws to allow an ever-increasing flow of nonwhites from the Third World. By *turning a blind eye to the illegal flow of millions more across our borders.* By *instituting programs* such as the Orwellian-sounding *affirmative action* and other forms of antiwhite job and education discrimination. By *taxing hard working, honest white workers and farmers* to support the growth of the criminal underclass in American's urban jungle. By twisting our social welfare laws which were supposed to help honest workers who were down on their luck into a scheme for promoting *the never ending maternity of millions of unmarried and unemployable, nonwhite welfare mothers,* while encouraging pregnant white girls to see the next generation of white babies as a problem which needs to be solved by the *abortionist's knife.* By *permitting* and then encouraging *interracial sex* through the use of change agents in our churches, in our schools and in our mass media. And, by the planned promotion of *homosexuality* and other forms of sexual perversion. By all these efforts and many more, the enemies of our people have worked for decades to destroy the America we loved. And always, whenever patriotic Americans have banded together to restore our nation, the *controlled media* and anti-American pressure groups like the foreign-controlled *ADL,* have tried to frighten away as many people as possible by shrinking and spitting and *calling the patriots anti-Semitic, neo-Nazi, racists,* etc., etc., etc., *ad infinitum, ad nauseam.*[70]

Here, the National Alliance knits together a host of typical Republican Party issues—the anti-immigration, anti-affirmative action, antitax, antiwelfare, antiabortion, antihomosexual agenda—with a yarn about racist conspiracy. In short, whether coded, semirational, or completely outrageous, all these conspiracy theories advance the same agenda: the reactionary, antidemocratic, quasi-fascist agenda.

Moreover, certain quasi-fascist ideals permeate the New Militia movement, such as a rejection of Communism, internationalism, and pacifism, as well as a hostility to many aspects of liberal democracy—in particular the ones that brought us a large federal government, the welfare state, environmentalism, civil rights legislation, affirmative action, and so on—all coupled with the valorization of revolutionary violence and heroism. Thus, although the New Militias use the traditional American rhetoric of the Citizen-Soldier ideal, the movement was founded by white supremacists and anti-Semites on the extreme Right, and their antidemocratic agenda still permeates the agenda of the larger movement.[71]

Speaking of the anti-immigration, anti-affirmative action, antitax, antiwelfare, antiabortion, antihomosexual agenda, Pat Buchanan's strong showing early in the 1996 presidential primaries illustrates the strength of the New Militia movement.[72] In fact, Buchanan explicitly targeted the New Militias in his campaign. Sounding many of the themes indicative of New Militia supporters, Buchanan proclaims the loss of U.S. sovereignty to internationalist organizations: "GATT . . . say[s] that . . . control over world trade is transferred from the Congress of the United States, where our own founding fathers placed responsibility, to a global institution in Geneva called a World Trade Organization, where . . . our American vote can be canceled out by Fidel Castro." Moreover, Buchanan believes the federal government has overstepped its bounds: "People ask me, what is our campaign all about, and could you sum it up in a couple of words? And, I said, my campaign is about restoring the Constitutional Republic. . . . In this Constitution, certain responsibilities and duties are given to the Federal Government and the rest belong to the states respectively, and to the people." Not surprisingly, in response to these supposed violations of the Constitution, Buchanan appeals to both identity politics and the Citizen-Soldier tradition. He explicitly claims to speak for the same angry white men mobilized by the New Militia movement—"unemployed, angry white male may seek Presidency," he quips. Moreover, Buchanan explicitly refers to the Citizen-Soldier tradition: "People say, What does sovereignty mean, Pat? I say go down to the village green in *Lexington*. Go down and stand there, where 17-year-old boys stopped the greatest army in the world."[73]

While these campaign themes might not sound so outrageous, Pat Buchanan, like the New Militia movement itself, is linked directly to his quasi-fascist supporters. That is, Buchanan has taken up the agenda of David Duke, who founded the National Association for the Advancement of White People, which was founded by the anti-Semitic Liberty Lobby in an attempt to unite white supremacists and neo-Nazis.[74] Duke and other far Right leaders align themselves with the Buchanan campaign—a fact that supposedly caused the candidate quite a bit of embarrassment.[75] Moreover, American quasi-fascist groups are quite proud of the fact that, "Vladimir Zhirinovsky . . . said he and Buchanan could work together to deport Jews from America and Russia. He called Buchanan a 'brother in arms.' "[76] So while Aryan Nations and other far Right groups arm themselves to defend white interests in the coming race war against ZOG (the Zionist Occupation Government)—the U.S. federal government—under the guise of the Citizens' Militias, Pat Buchanan, heir of the racist Right, advocates a reactionary identity politics for angry white males in the electoral arena and has found quite a few supporters.

Performing White Masculinity

What we have discussed so far is not the whole story. As we recall from previous chapters, the Citizen-Soldier ideal traditionally played a key role in the construction of masculinity. Thus, it is no accident that economically threatened, angry white men have chosen to deal with their problems by engaging together in the martial practices traditionally constitutive of *armed masculinity*. Consider that many of these angry white men are economically threatened and thus unable to provide adequately for their families. One could say they have been rendered economically impotent—emasculated. Thus, they find the practices traditionally productive of masculinity appealing. In other words, they are, in essence, reconstituting themselves as masculine subjects by bearing arms in the New Militias.

In fact, the New Militia movement can be put in the context of a broad-scale "remasculinization of America" after the humiliating loss in Vietnam—the emasculation of America—not to mention the full-force assault on white masculinity perpetrated by newly politicized movements of blacks, women, lesbians, and gays. Susan Jeffords argues that after the war, "the male Vietnam veteran—primarily the white male—was used as an emblem for a fallen and emasculated American male, one who had been falsely scorned by society and unjustly victimized by his own government."[77] That is, in reaction to the "new social movements" of the 1960s and 1970s, the American Right began to use the image of the Vietnam vet in its effort to create an identity politics for angry white men. Through the deployment of this image,

> masculinity [began] to place itself in the category of a social group in need of special consideration. No longer the oppressor, men came to be seen, primarily through the imagery of the Vietnam veteran, as themselves oppressed. It was not then difficult to insert this characterization into an already formulated cultural attitude toward the victimized that had been established in relation to civil rights and women's movements, to the point that hiring quotas and organizations like NOW were seen as depriving men of their "rights."[78]

Indeed, Vietnam vets form a key constituency of the New Militia and its reactionary identity politics for angry white men.

But more than just constituting a new identity politics, the deployment of the image of the victimized Vietnam vet played a central role in right-wing attacks on the federal government. As Jeffords explains it, "the accumulated negative features of the feminine" were transferred "to the government itself, the primary vehicle for legislated and enforced changes in civil rights. From this vantage point, not only could individual men cite discrimination (Jim Baake), but all men as a group could also declare their suffering at the hands of a government biased toward and operating under the aegis of the feminine."[79] Consequently, the image of the Vietnam vet

> became the springboard for a general remasculinization of American culture that is evidenced in the popularity of figures like Ronald Reagan, Oliver North, and J. R. Ewing, men who show open disregard for government legislation and legal decisions and favor images of strength and firmness with an independence that smacks of Rambo and confirms their faith in a separate culture based on a mythos of masculinity.[80]

Moreover, the image helped lay the groundwork for the eventual emergence of the New Militia movement.

Participants in the New Militia movement engage in martial practices in an effort to reconstitute a white masculinity that is threatened by progressive democratic movements and by a government that is quickly becoming "female"—not to mention "black [and] brown." In fact, some militia supporters actually make direct references to masculinity in their rhetoric. For example, Identity Christian, Klansman, and neo-Nazi author David Lane asserts that the Jews have used the mass media "to insult and *emasculate the White man* while depicting nonwhite males to be heroes so White women would desert their Race by the millions."[81] Aryan Nations leader Richard Butler states: "*You don't have anybody today who is a man,* who has stood up for his race."[82] And Bob Mathews, founder of the Order, modeled after the group portrayed in the neo-Nazi, white supremacist book, *The Turner Diaries,* said, "when it became obvious that we were going to do more than just talk,"—that is engage in murder and bank robberies—"most of the men started backing out and turning their backs to *those of us who have retained our manhood* and our Aryan pride. . . . I know what future awaits our children *unless I stand up like a man and fight.*"[83] Alternately, he states, "I have no choice. I must *stand up like a White man and do battle.*"[84] Clearly for Mathews, being a man requires action: He must "stand up like a man and fight." And again, at a neo-Nazi National Alliance convention, Mathews reiterated his point: "So stand up *like men,* and drive the enemy into the sea! Stand up *like men,* and swear . . . that you will reclaim what *our* forefathers discovered, explored, conquered, settled, built and died for! Stand up *like men* and reclaim our soil!"[85] Mathews founded a paramilitary group to advance the interests of Aryan men.

His phrasing is interesting because it implies that to *be* a man, one must *act like* a man—the point I have been making throughout this study. As Judith Butler perceptively puts it, "[T]he articulation 'I feel like a woman' by a female or 'I feel like a man' by a male presupposes that in neither case is the claim meaninglessly redundant. Although it might appear unproblematic *to be* a given anatomy, . . . the experience of a gendered psychic disposition or cultural identity is considered an achievement."[86] In other words, since gender is performatively constructed rather than rooted in nature, there is no essential masculinity outside of that constructed through engagement in masculine practices. To be a man, one must constantly *act like* a man.

Traditionally, martial practices have functioned to construct a masculinity defined in direct opposition to femininity and to keep the feminine threat inside men at bay. Hannah Pitkin emphasizes this in her discussion of Machiavelli's stress on military service in the militia: "Only ferocious discipline and terrifying punishments" can prevent men from becoming feminine.[87] Linda Zerilli argues that contemporary man fears the breakdown of gender identity; he fears that "if the code of gender difference is not strictly adhered to at each and every moment, all is lost."[88] Klaus Theweleit finds an extreme version of the same phenomenon in his examination of the erotic writings of the Freikorpsmen who became Nazi SA officers: "What fascism promised men was . . . dominance of the hostile 'female' element within themselves."[89] In the introduction to his second volume, Jessica Benjamin and Anson Rabinbach tell us that Theweleit criticizes the traditional Frankfurt School analysis of fascism for neglecting the "attraction of fascism itself"—its "passionate celebration of violence."[90] They then go on to summarize his argument as follows: "Indeed, it is Theweleit's insistence on the primacy of violence—originating in the fear and hatred of the feminine—that distinguishes his approach from the older social-psychological models.... The crucial element of fascism is its explicit sexual language.... this fascist symbolization creates a particular kind of psychic economy which places sexuality in the service of destruction."[91] Martial practices produce an *armed masculinity* constituted in direct opposition to femininity, and fascism simply exaggerates this process.

To the extent that the New Militia incorporates fascistic elements into its worldview, using Theweleit's research makes sense. What I want to underline here is that pleasure exists in military discipline and violence, a point we will discuss in more detail in the next chapter. In fact, Nancy Hartsock argues that in Western culture "hostility and domination . . . are central to sexual excitement. . . . What *is* sexually exciting in Western culture is hostility, violence and domination, especially but not necessarily directed against women."[92] We should not overlook the pleasure involved in martial practices. As William James notes, war making is a "thrill."[93] And running around in the woods wearing camouflage and shooting guns is a whole lot of fun as well.

The precariousness of performatively constructed masculinity helps explain the urgency of the right-wing man's opposition to feminism, lesbian and gay rights, and even the welfare state. The militiamen stand opposed to women's struggle to control their own reproductive capacities, to welfare benefits that provide subsistence for women who choose to live outside the bounds of marriage, and of course to legal protections for lesbians and gay men[94] because these liberal agenda items undercut traditional configurations of gender.[95] As the National Alliance puts it, "we see the products of this system all around us: too many weak, indecisive men and too many unfeminine women."[96] Only the denial of equal rights to gay men and all women, the denial to them of full citizenship, or, even better, the purging of femininity and homosexuality from the public sphere and the military—the two traditional realms of citizenship—helps secure an always-unstable masculinity. Thus, one of the phenomena fueling the New Militia movement is the instability of gender in an age of feminism, lesbian/gay rights, and a real or imagined economic impotence among white men.

Citizen-Soldiers or Blood Brothers?:
Evaluating the New Militia Movement

The New Militia movement constitutes an increasingly successful effort by quasi-fascist fringe groups to create a coalition among very different groups of angry white men. Using the Second Amendment as a springboard, the New

Militia movement has been able to lay claim to the Citizen-Soldier tradition in America, and clearly their rhetoric speaks to many Americans. But why? Why has the New Militia movement been so successful in their appeals to the American public? What does the Citizen-Soldier ideal signify to people?

Apparently, the Citizen-Soldier ideal remains firmly entrenched in American political mythology. As we saw in chapter 4, the Citizen-Soldier ideal used to play a key role in the constitution of masculine republican citizens. Although the Citizen-Soldier in its classic form essentially died during the combat of the nineteenth century, for some reason the ideal still resonates with many white American men. The Citizen-Soldier ideal constitutes a powerful symbol that speaks to the "widespread political anxieties, traumatic experiences, and unfulfilled wishes" of a white male America in crisis—for five reasons.[97]

First of all, the ideal of the Citizen-Soldier remains in American political memory as the symbol of participatory citizenship. In a time of seemingly unprecedented change and conflict, people want to have some control over the institutions that are supposed to serve their interests. Yet all around them, citizens from across the political spectrum see decisions being made with which they do not agree, and they feel powerless to do anything about it. How did the head of the Haitian secret police accidentally slip into the United States? How could a jury acquit the police officers who beat Rodney King? Is police brutality so routine that it is reasonable for an innocent man to run from the police in certain areas of town? Did the CIA deliberately introduce crack cocaine into inner-city neighborhoods? Was O. J. Simpson framed by the Los Angeles Police Department? How could jurors acquit O. J. after he brutally killed his ex-wife? Why is the government funding studies to see how fast ketchup runs? Why are the schools distributing condoms? Why are our children reading *Heather Has Two Mommies*? Why is the government putting tracking devices in our money? Did the U.S. Navy really shoot down TWA flight 800? Many people no longer feel that their democratic institutions are accountable to the citizenry, and they feel powerless to do anything about it. The Citizen-Soldier ideal appeals to many people because it represents an active form of popular sovereignty that will force our ostensibly democratic institutions to act responsibly and correctly.

Second, the Citizen-Soldier ideal appeals to people because it signifies the idea of legitimacy, during the current "crisis of legitimacy" we are experiencing as a nation. Many people believe they are being subjected to laws and policies to which they never consented. And because they feel shut out of the political process, they feel they have no recourse. Within American political mythology, the Citizen-Soldier represents the resistance of the American people to illegitimate rule, and that is why a lot of Americans are joining the New Militia movement.

Third, the Citizen-Soldier ideal recalls a time when American society supposedly had a consensus about basic values, when it was not plagued with the conflicts engendered by multiculturalism and "identity politics." However, the

Citizen-Soldier ideal also signifies a time when full citizenship was restricted to white men only. Thus, it is no accident that the Citizen-Soldier ideal appeals particularly to white men who feel threatened by the democratic reforms encapsulated by the term *multiculturalism.*

Fourth, the Citizen-Soldier embodies traditional gender norms—from a time when men were soldiers fighting to protect wives and children and were also citizens governing in the best interests of their family and community. This vision of the proper ordering of society accompanies nicely the myth of the patriarchal nuclear family with the male breadwinner and the female housewife and stay-home mother. In short, the Citizen-Soldier ideal harks back to a time when "men were men and women were women." Consequently, it appeals to more traditional Americans who see their way of life threatened by feminism and the lesbian/gay civil rights movement as well as by the state that they believe serves those interests. And finally, the Citizen-Soldier symbolizes potent masculinity in a time where this ideal is threatened from a variety of angles.

Calling themselves "patriots," members of the New Militia movement claim to stand within the civic republican heritage of the American Revolution—the Citizen-Soldier tradition. "The patriots of two centuries ago fought a government in London that they believed imposed unreasonable taxes, trampled their rights, and was unresponsive to their needs. Today's self-styled patriots shake their fists at the government in Washington, D.C., which they believe imposes unreasonable taxes, tramples on individual freedom, and is unresponsive to their needs."[98] But are these two movements comparable?

How does the New Militia Movement stand vis-à-vis the Citizen-Soldier tradition? On one hand, the New Militias follow directly in the footsteps of the nineteenth-century militia tradition at its most vicious. For example, the New Militia movement resuscitates the tradition of the white supremacist militias of the Old South, discussed in chapter 4. Indeed, the New Militias offer the traditional vices of the Citizen-Soldier tradition: xenophobia, racism, violence, and homogeneity. However, as in the nineteenth century, these vices are interconnected with a corresponding set of virtues. So while the New Militias do indeed offer xenophobia, racism, violence, and homogeneity, at the same time they offer patriotism, fraternity, devotion to a common good, and a common identity—to those they include in their membership.

Members of the racist Right come to think of themselves as white people—take on the identity of white people—as they engage together in communal practices, such as preparing for the race war and worshiping at Christian Identity churches. The festivals and rituals of the Right function in a way similar to the civic practices of early republican America, in terms of their role in the constitution of identities. However, the results are very different: Aryan Christian practices produce white supremacists and anti-Semites rather than republican citizens. Consider the following example:

The sounds of children at play rose from the swing set and jungle gym at the base of the guard tower, while the adults listened to calls for war against the Jews. Many of the women felt comforted by talk of the men laying down their lives to protect their families, and values. Some of the women cut locks of their hair, tied them with ribbons, and presented them to their men, so that they could carry them into battle. . . . [The Aryan Nations camp] took on a jamboree flavor during [Butler's] congresses. The grounds were covered with tents, trailers, and pickups with camper shells, and people walked among them to socialize.

Men wore blue uniforms and bore arms.[99] These individuals came to think of themselves as Aryan Christians and to love their community, as they participated together in the Aryan Nations congress. In short, engagement in Aryan practices produces Aryan identities.

So in keeping with the Citizen-Soldier tradition, the New Militia movement offers a love of community instilled through participation with others in common actions. It offers its members the chance to be included in something larger than themselves. It offers fraternity to those who are allowed to participate — a blood brotherhood for Aryan Christian men. And the New Militia movement offers economically threatened white males a masculinity revitalized through engagement in martial practices. Once again, we can see that the virtues and the vices of the Citizen-Soldier tradition are inextricably linked. The New Militias and the Aryan Christian Right call upon the vicious, undemocratic aspects of the Citizen-Soldier tradition and in so doing also reap some of the virtues of the tradition. They are creating fraternity along with racism, patriotism along with xenophobia, and community action along with violence.

On the other hand, however, the Aryan Christian Militia movement violates three basic principles of the *political theory* of the Citizen-Soldier. First, while the Aryan Christian movement embraces martial practices, it does not view them as necessarily wedded to democratic republican ideals such as liberty, equality, the rule of law, and participatory citizenship. It does not accept the *Citizen* half of the Citizen-Soldier ideal, and the tradition of civic republicanism requires that these two halves be linked. In this sense, Aryan Warriors are not Citizen-Soldiers.

Second, while the New Militias pay a lot of lip service to the traditions of the American Founding, the Aryan Christian movement accepts the Declaration of Independence, the Articles of Confederation, the American Constitution, and the Bill of Rights not as civic republican or even liberal documents, but as the revealed word of God.[100] This violates the secular nature of the Citizen-Soldier tradition that emerged in direct opposition to a Christian worldview. As noted in chapter 2, in its fifteenth-century origins the political movement of civic republicanism constituted an attack on the medieval Christian worldview, with its traditional, hierarchic view of society. Then in the eighteenth century, as noted in chapter 3, civic republicanism again forged itself in opposition to religiously justified monarchical and aristocratic rule. Citizenship requires liberty rather than subjection to tradition, equality rather than hierarchy and rank, fraternity rather than paternity and filiality, and autonomy rather than obedience to natural, God-given law and dependence on natural superiors. Civic republicanism requires that citizens govern themselves for the common good through the rule of *man-made*—not Divine—law. It cannot accommodate the idea of a revealed Truth that citizens must not question.

Finally, the Aryan Christian militia vision of the far Right violates the political theory of the Citizen-Soldier, because it rejects the *citizenship of civic practices* for a *citizenship of blood*. For example, Aryan Nations leader "Butler told [his] assemblage that granting citizenship to non-Christians and nonwhites was part of a Zionist plot to adulterate Aryan purity."[101] Although the civic republican tradition may have *actually* included only a homogeneous group of men, it explicitly rejects on *principle* the idea that citizenship should be extended only to people with particular bloodlines—such as Aryan blood. Civic republicanism entails a commitment to universalizable principles. Though these principles may be violated in *practice,* they cannot be violated in *principle.*

Thus, although the New Militia movement does take up the legacy of the Citizen-Soldier tradition in the most vicious sense, it cannot legitimately claim to stand within the civic republican tradition, because it violates the democratic principles in which the tradition is theoretically rooted. In other words, the New Militia movement does in fact resemble those nineteenth-century militias that in *practice* forged their identities in direct opposition to denigrated "others" and mobilized for the purposes of shoring up white supremacy. However, *such exclusionary practices even then directly violated the political theory of the Citizen-Soldier,* as I argued in chapter 4. Thus, while exclusionary practices do form a central part of the legacy of the Citizen-Soldier tradition in America, they represent that tradition at its most vicious and antidemocratic.

As I have argued throughout the book, it is my thesis that the political theory of the Citizen-Soldier constitutes a normative tradition with a commitment to universalizable democratic principles, including liberty, equality, camaraderie, the rule of law, the common good, civic virtue, and participatory citizenship. Therefore, if we consider the *political theory* of the Citizen-Soldier in its best, *democratic* sense, then we must reject the claim of antidemocratic groups that they stand within that civic republican tradition. Thus, as long as the New Militia movement works to advance the antidemocratic agenda of the quasi-fascist Right, we do not have to recognize the movement as the legitimate heirs of the democratic tradition of the Citizen-Soldier in America.

Notes

1. Here I make reference to the terminology of F. Plasser and P. Ulram, who have discussed the "politics of masks, symbols, and emotions" by fascist groups in Europe. Fritz Plasser and Peter Ulram, "Wahltag ist Zahltag: Populistischer Appell und Wahler-

protest in den achzinger Jahren," *Osterreichische Zeitschrift fur Politikwissenschaft,* 89.2 (1989): 161. I am indebted to Manfred Steger for a translation of these ideas into English (personal conversations during January 1999) and for his discussion of these ideas in the Austrian context in "The 'New Austria,' the 'New Europe,' and the New Nationalism," paper presented at the annual meeting of the American Political Science Association, Washington, DC, September 1993.

2. For a discussion of the "culture wars," see James Davison Hunter, *Culture Wars: The Struggle to Define America* (New York: Basic Books, 1992).

3. Amy Gutmann and Dennis Thompson, *Democracy and Disagreement* (Cambridge: Belknap Press of Harvard University Press, 1996), 1.

4. For a discussion the widespread "disconnect" between citizens and democratic institutions, see Joseph S. Nye, David C. King, and Philip D. Zelikow, eds., *Why People Don't Trust Government* (Cambridge: Harvard University Press, 1997). For a recent empirical study, see Doble Research Associates, Inc., *Governing America: Our Choices, Our Challenge, A Report on 1997–98 National Issues Forums: How People Are Thinking About Democratic Government in the U.S.,* available from National Issues Forums Research, 200 Commons Road, Dayton, OH 45459-2799.

5. Sternhell differentiates Nazism from fascism: "The basis of Nazism was racism in its most extreme sense, and the fight against the Jews, against 'inferior races,' played a more preponderant role in it than the struggle against communism. . . . In fact, racial determinism was not present in all varieties of fascism. . . . Racism was thus not a necessary condition for the existence of fascism." Zeev Sternhell with Mario Sznajder and Maia Asheri, *The Birth of Fascist Ideology: From Cultural Rebellion to Political Revolution,* trans. David Maisel (Princeton: Princeton University Press, 1994), 4–5. For an opposing view see Ernst Nolte, *The Three Faces of Fascism,* tr. Leila Vennewitz (New York: Holt, Rinehart and Winston, 1966).

6. Sternhell argues that two key characteristics of fascism are "on the one hand, a rejection of democracy, Marxism, liberalism, the so-called bourgeois values, the eighteenth century heritage, internationalism, and pacifism, on the other hand, a cult of heroism, vitalism, and violence." See *Birth of Fascist Ideology,* 32.

7. See John Whiteclay Chambers II, *Tyranny of Change: America in the Progressive Era 1890–1920,* 2nd ed. (New York: St. Martin's Press, 1992).

8. Victor Ostrovsky and Claire Hoy, "By Way of Deception Thou Shalt Do War," *Race and Reason* 1 (Jan./Feb. 1993): 10, emphasis mine. Originally published in *National Vanguard Magazine,* P.O. Box 330, Hillsboro, WV 24946. As of May 1999 available at http://www.stormfront.org.

9. National Alliance, "National Alliance Goals," National Alliance Main Page, 1996, emphasis mine.

10. National Vanguard Books staff, *Who Rules America?* (Hillsboro, WV: National Vanguard Books, 1993), 3, emphasis mine.

11. American Dissident Voices, "ADL: America's Greatest Enemy," ADV Directory//National Alliance Main Page, Program of 29 May 1993, emphasis mine.

12. Steger, " 'New Austria,' " 8.

13. National Alliance poster, emphasis mine.

14. Kevin Flynn and Gary Gerhardt, *The Silent Brotherhood: The Chilling Inside Story of America's Violent Anti-Government Militia Movement* (New York: Penguin Books, 1995), 75–76.

15. Flynn and Gerhardt, *Silent Brotherhood,* 7.

16. National Alliance, "What Is the National Alliance?" National Alliance Main Page, 1995.

17. For a discussion of the distinction between *ius solis* and *ius sanguinis* in the German context, see Manfred Steger and F. Peter Wagner, "Political Asylum, Immigration, and Citizenship in the Federal Republic of Germany," *New Political Science* 24/25 (Summer 1993): 65.

18. National Alliance, "The Saga of White Will," *New World Order Comix* no. 1 (Hillsboro, WV: National Vanguard Books, 1993).

19. National Alliance, "National Alliance Goals."

20. Flynn and Gerhardt, *Silent Brotherhood,* 23.

21. Flynn and Gerhardt, *Silent Brotherhood,* 80.

22. Flynn and Gerhardt, *Silent Brotherhood,* 23.

23. Flynn and Gerhardt, *Silent Brotherhood,* 22.

24. "Demographics and Revolt," http://www.io.com/wlp/aryan-page/y04.html., 2/17/96, 1.

25. "Demographics and Revolt," 1.

26. This is a well-documented fact. For example, see Flynn and Gerhardt, *Silent Brotherhood;* Jill Smolowe, "Enemies of the State," *Time,* 8 May 1995, 58–65; and Joseph P. Shapiro, "An Epidemic of Fear and Loathing: Bar Codes, Black Helicopters and Martial Law," *U.S. News and World Report,* 8 May 1995, 37–44.

27. Jonathan Karl, *The Right to Bear Arms: The Rise of America's New Militias* (New York: HarperPaperbacks, 1995), 34.

28. Karl, *Right to Bear Arms,* 44.

29. James Coates, *Armed and Dangerous: The Rise of the Survivalist Right* (New York: Hill and Wang, 1995), 110.

30. Robert B. DePugh, "Political Platform of the Patriotic Party," in *Extremism in America,* ed. Lyman Tower Sargent (New York: New York University Press, 1995), 95.

31. Coates, *Armed and Dangerous,* 110.

32. "This nation should immediately withdraw from the United Nations." DePugh, "Political Platform," 101.

33. See Coates, *Armed and Dangerous,* 110, and Karl, *Right to Bear Arms,* 114.

34. David Dawson, "Posse Comitatus," in *The Military Draft: Selected Readings on Conscription,* ed. Martin Anderson with Barbara Honegger (Stanford, CA: Hoover Institution Press, 1982), 5.

35. The Posse Comitatus, "It Is the Duty of Government to Prevent Injustice—Not to Promote It," in *Extremism in America,* ed. Lyman Tower Sargent (New York: New York University Press, 1995), 345–46, emphasis mine.

36. Posse Comitatus, "Duty of Government," 346.

37. Posse Comitatus, "Duty of Government," 344.

38. Posse Comitatus, "Duty of Government," 345.

39. See Coates, *Armed and Dangerous,* 110; Karl, *Right to Bear Arms,* 114.

40. Marc Cooper, "A Visit with MOM: Montana's Mother of All Militias," *The Nation,* 22 May 1995, 722.

41. Karl, *Right to Bear Arms,* 115.

42. Cooper, "A Visit with MOM," 716.

43. "Old List of Klan Members Recalls Racist Past in an Indiana City," *New York Times,* 2 August 1995.

44. For an example of the Jewish banking conspiracy, see Egon Caesar Corti, *The Rise of the House of Rothschild* (Chattanooga, TN: Western Islands, 1972).

45. Steger, "New Austria," 8.

46. Coates, *Armed and Dangerous,* 83–84.

47. Coates, *Armed and Dangerous,* 14.

48. Karl, *Right to Bear Arms,* 115, emphasis mine.

49. See "American Program," in *Extremism in America,* ed. Lyman Tower Sargent (New York: New York University Press, 1995), 132.

50. See "The Populist Party of the United States Platform 1984," in *Extremism in America,* ed. Lyman Tower Sargent (New York: New York University Press, 1995), 18–25. In short, the Populist Party wants to restrict immigration, abolish "affirmative action" and racial quotas, reform the welfare system, crack down on crime, reduce federal income taxes, protect American farmers and workers, place America's interests first, reform Wall Street, establish the strongest defense in the World, and bring down interest rates.

51. Smolowe, "Enemies of the State."

52. "Home-Grown Courts Spring Up as Judicial Arm of the Far Right," *New York Times,* 17 April 1996.

53. "In Montana and throughout the country the militias forged a new alliance between the old-line white supremacist groups and the newer antitax organizations, property rights organizations and Wise Use anti-enviro activists . . . , Christian conservatives, antiabortion militants, Perotista constitutionalists, gun-owner associations and thousands of individual representatives of that newly categorized political species, the Angry White Male." Cooper, "A Visit with MOM," 718–19.

54. Cooper, "A Visit with MOM, 714–22, 716.

55. Sara Diamond, *Roads to Dominion: Right-Wing Movements and Political Power in the United States* (New York: Guilford Press, 1995), 257–58.

56. Diamond, *Roads to Dominion,* 13.

57. Diamond, *Roads to Dominion,* 257–58.

58. Coates, *Armed and Dangerous,* 13.

59. Karl, *Right to Bear Arms,* 109. Interestingly, Karl cites the 1991 Kettering Foundation study as evidence for the political discontent of the public.

60. Sternhell, *Neither Right Nor Left: Fascist Ideology in France* (Berkeley, CA: University of California Press, 1986), 1.

61. Karl, *Right to Bear Arms,* 53; and Coates, *Armed and Dangerous,* 83–84.

62. Karl, *Right to Bear Arms,* 57.

63. Smolowe, "Enemies of the State," 66.

64. "Bomb Suspect Felt at Home Riding the Gun-Show Circuit," *New York Times,* 5 July 1995.

65. Karl, *Right to Bear Arms,* 44.

66. Karl, *Right to Bear Arms,* 69.

67. Lyle Stuart recently made the controversial decision to publish this once hard-to-get text, along with an interesting preface asserting his disgust with the book, his personal antiracism, his own nonthreatened heterosexuality, and his guttural opposition to censorship. Andrew Macdonald, *The Turner Diaries* (New York: Barricade Books, 1996).

68. Coates, *Armed and Dangerous,* 9.

69. Coates, *Armed and Dangerous,* vii.

70. National Alliance and National Vanguard Books phone message, (216) 846–1045, 25 October 1993, emphasis mine.

71. Cooper, "A Visit with MOM," 714–22, 716.

72. For a more recent discussion of Buchanan's beliefs, see "With Great Gusto, Pat Buchanan Launches 3rd Run for Presidency," *Chicago Tribune,* 3 March 1999.

73. "Excerpts From Buchanan Campaign Speech," *New York Times,* 22 February 1996, emphasis mine.

74. Diamond, *Roads to Dominion,* 273, 261.

75. "Buchanan Drawing Extremist Support, and Problems, Too," *New York Times,* 23 February 1996.

76. "Thunderbolts," Stormfront White Nationalist Resource Web Page, 1996. See also "Russian Backer for Buchanan," *New York Times,* 23 February 1996.

77. Susan Jeffords, *The Remasculinization of America: Gender and the Vietnam War* (Indianapolis: Indiana University Press, 1989), 168–69.

78. Jeffords, *Remasculinization of America,* 169.

79. Jeffords, *Remasculinization of America,* 169.

80. Jeffords, *Remasculinization of America,* 169.

81. Quoted in Coates, *Armed and Dangerous,* 89.

82. Quoted in Flynn and Gerhardt, *Silent Brotherhood,* 77.

83. Flynn and Gerhardt, *Silent Brotherhood,* 153, emphasis mine.

84. Quoted in Coates, *Armed and Dangerous,* 52.

85. Quoted in Flynn and Gerhardt, *Silent Brotherhood,* 122.

86. Judith Butler, *Gender Trouble: Feminism and the Subversion of Identity* (New York: Routledge, 1990), 22.

87. Hannah Fenichel Pitkin, *Fortune Is a Woman: Gender and Politics in the Thought of Niccolo Machiavelli* (Berkeley: University of California Press, 1984), 136.

88. Linda Zerilli, *Signifying Woman: Culture and Chaos in Rousseau, Burke, and Mill* (Ithaca: Cornell University Press, 1994), 18.

89. Klaus Theweleit, *Male Fantasies,* vol. 1, *Women, Floods, Bodies, History,* trans. Stephen Conway in collaboration with Erica Carter and Chris Turner (Minneapolis: University of Minnesota Press, 1987), 434.

90. From the German edition, not included in the English edition, 534.

91. Klaus Theweleit, *Male Fantasies,* vol. 2, *Male Bodies: Psychoanalyzing the White Terror,* trans. Stephen Conway in collaboration with Erica Carter and Chris Turner (Minneapolis: University of Minnesota Press, 1989), xii.

92. Nancy Hartsock, *Money, Sex, and Power: Toward a Feminist Historical Materialism* (Boston: Northeastern University Press, 1983), 157, 166.

93. William James, "The Moral Equivalent of War," in *Education for Democracy,* ed. Benjamin R. Barber and Richard Battistoni (Dubuque, IO: Kendall/Hunt, 1993), 94.

94. See Coates, *Armed and Dangerous.*

95. On 26 February 1996, the *New York Times* reported that members of right-wing paramilitary groups rallied behind the cause of a doctor who engaged in a six-day standoff with the F.B.I. after failing to appear at a hearing on charges of failing to pay $70,000 in back child support. Here again, the militia groups oppose measures that would aid women who choose to live outside the bounds of marriage.

96. National Alliance, "National Alliance Goals."

97. Steger, "New Austria," 8.

98. Karl, *Right to Bear Arms*, 3–4.

99. Flynn and Gerhardt, *Silent Brotherhood*, 116–17.

100. According to Christian Identity theology, "after settling the New World, the true Promised Land, America's founding fathers were inspired by God to write the sacred documents we know as the Declaration of Independence, the Constitution, and the Bill of Rights. The amendments that followed, according to Identity, are Satanic additions dictated through today's Jews to undermine the white race." Flynn and Gerhardt, *Silent Brotherhood*, 72. Also see Coates, *Armed and Dangerous*.

101. Flynn and Gerhardt, *Silent Brotherhood*, 83.

Chapter 6

Troubling *Armed Masculinity:*
Military Academies, Hazing Rituals,
and the Reconstitution of the Citizen-Soldier

One of Virginia's educational institutions is military in character. Are women to be admitted on an equal basis, and, if so, are they to wear uniforms and be taught to bear arms?

— Judge of the Virginia courts, 1970

We know how to train young men to be men. We don't know how to train young women to be men.

— Public relations director for The Citadel, 1994

The Citizen-Soldier tradition places military service at the center of its vision of civic education. According to traditional arguments, military service teaches individuals the virtues necessary for republican citizenship: selflessness, courage, fraternity, patriotism, and civic virtue—the willingness to put the common good ahead of individual self-interest, including sacrificing one's life, if necessary. The Citizen-Soldier represents a model of citizenship in which these martial virtues become attached to citizenship. Historically, this has meant that only men could become republican citizens, because only men were allowed to be soldiers. But now that women are supposed to be full citizens, we have to ask: What happens, in a tradition that links citizenship to soldiering, when women become citizens? If military service forms a central part of re-publican citizenship, then if women want to become republican citizens, they must engage in military service alongside men. And indeed, at various points in history, women have fought for the right to bear arms in defense of the re-public and have actually served in the military disguised as men.[1] These women saw military service as essential to their full citizenship.

Performativity Theory and the Possibility
of *Subversive Transgender Performances*

Can women become citizen-soldiers alongside men if they engage in the same set of civic and martial practices? The *performativity theory* of identity, as I have used it throughout this study, leaves open the possibility that we can radically reconstruct gender and citizenship by altering the practices in which we engage. As I have demonstrated in my discussion of Machiavelli and Rousseau, as well as my discussion of American political history, within the civic republican tradition masculinity is not understood as a characteristic occurring naturally in biological males. To the contrary, only by engaging in traditionally masculine practices, such as soldiering, do biological males become culturally masculine. Likewise, citizenship for civic republicanism is not simply a category bestowed upon anyone born within a particular bounded territory (as with *ius solis*) or born into a particular ethnic group (as with *ius sanguinis*). To the contrary, citizenship is constituted by participation in particular civic and martial practices. Furthermore, within this tradition, citizenship and masculinity are produced *simultaneously* by the repeated engagement in the same set of civic and martial practices. In short, one is not born but rather *becomes* a masculine citizen, as he performs those tasks required by the prescriptive ideal of the Citizen-Soldier. Thus, if identity is formed through engagement in particular practices, then we should be able to change our identities by changing the practices in which we engage.

My understanding of civic and gender identity as performatively constructed creates important opportunities for democratic and feminist theorists interested in creating a more just society. That is to say, if gender is performatively constructed, rather than rooted in nature, this means that gender identity is malleable rather than fixed. Because gender is constructed, it can also be reconstructed in a way that does not advantage one particular gender over another. Thus, if masculine citizen-soldiers are traditionally constituted through engagement in civic and martial practices, and feminine subjects are traditionally constituted through the exclusion from these same practices, then the engagement of "women" in the practices constitutive of masculine republican citizenship should allow them to become citizen-soldiers on the same basis as men. That is, the *subversive transgender performances* of "women" acting out "male" scripts could work to highlight the artificiality of normative constructions of gender and consequently undermine the sexism such constructions generate.

The concept of performativity I am using in this work significantly reinterprets Judith Butler's original notion of performativity as laid out in her seminal work, *Gender Trouble*. Butler begins as follows:

> If gender is the cultural meanings that the sexed body assumes, then a gender cannot be said to follow from a sex in any one way. Taken to its logical limit, the

sex/gender distinction suggests a radical discontinuity between sexed bodies and culturally constructed genders. . . . The presumption of a binary gender system implicitly retains the belief in a mimetic relation of gender to sex whereby gender mirrors sex or is otherwise restricted by it. When the constructed status of gender is theorized as radically independent of sex, gender itself becomes a free-floating artifice, with the consequence that *man* and *masculine* might just as easily signify a female body as a male one, and *woman* and *feminine* a male body as easily as a female one.[2]

If gender is entirely a social construct, with no fixed referent in nature, then there is no reason why only biological males can become *masculine* and only biological females *feminine*.

Making this logical step allows Butler to develop her immensely important concept of performativity. Building on Simone de Beauvoir's famous claim that "one is not born, but rather *becomes* a woman," Butler argues that "*woman* itself is a term in process, a becoming, a constructing that cannot rightfully be said to originate or to end." That is to say, to argue that the process of becoming gendered could ever be finally finished would be to imply that "a *telos* . . . governs the process of acculturation and construction," which would be to reinscribe a natural foundation for gender identity. Maintaining the integrity of the sex/gender split, Butler argues for the performative construction of gender: "Gender is the repeated stylization of the body, a set of repeated acts . . . that congeal over time to produce the appearance of substance, of a natural sort of being. . . . No longer believable as an interior 'truth' of dispositions and identity, sex will be shown to be a performatively enacted signification (and hence not 'to be')."[3] That is, what *appears* to be gender identity must continually be produced and reproduced through the performance of gender-appropriate behaviors. "Men" and "women" are constantly becoming gendered as they participate in behaviors required by cultural norms of masculinity and femininity.

Given this notion of *performativity,* Butler seems to suggest the possibility of actively changing cultural constructions of gender. She says she wants

to think through the possibility of *subverting and displacing those naturalized and reified notions of gender that support masculine hegemony and heterosexist power, to make gender trouble,* not through the strategies that figure a utopian beyond, but *through the mobilization, subversive confusion, and proliferation of precisely those constitutive categories that seek to keep gender in its place by posturing as the foundational illusions of identity.*[4]

This passage seems to suggest that "men" and "women" could actually subvert hegemonic norms of masculinity and femininity by transgressively engaging in counterhegemonic gender behavior.

Nevertheless, while Butler seems to hint at the possibility of radically reconstructing gender in the preceding passages, she quickly retreats from such

constructive possibilities in her subsequent work, *Bodies That Matter*. Attempting to clarify "questions raised by the notion of gender performativity introduced in *Gender Trouble*," Butler explains that,

> if I were to argue that genders are performative, that could mean that I thought that
> one woke in the morning, perused the closet or some more open space for the gen-
> der of choice, donned that gender for the day, and then restored the garment to its
> place at night. Such a willful and instrumental subject, one who decides *on* its gen-
> der, is clearly not its gender from the start and fails to realize that its existence is
> already decided *by* gender. Certainly, such a theory would restore a figure of a
> choosing subject—humanist—at the center of a project whose emphasis on con-
> struction seems to be quite opposed to such a notion.[5]

Though agreeing with Butler that choosing a gender is not as simple and voluntaristic as choosing to don a skirt or fatigues, I have much fewer reservations
about placing the notion of performativity within a humanistic political framework. Indeed, I argue for the importance of restoring to the debates about the
construction of gender a stronger sense of agency than is possible within Butler's postmodern paradigm.[6]

In spite of her theoretical agenda, Butler paradoxically insists that performativity actually entails a lack of freedom. According to her definition, "performativity is neither free play nor theatrical self-presentation; nor can it be
simply equated with performance." Instead, performativity entails "the forced
reiteration of norms" within a matrix of constraints.[7] Moreover, the subject
cannot perform gender as he or she chooses because the humanist notion of the
"choosing subject" is merely a fiction: "repetition is not performed *by* a subject; [rather] this repetition is what enables a subject and constitutes the temporal condition for the subject"; there is no "doer behind the deed."[8]

In other words, Butler's postmodern commitments ironically truncate the
very possibility of a radical reconstruction of gender that her earlier work seems
to suggest, because her rejection of a "doer behind the deed" eliminates the possibility of meaningful agency. Seyla Benhabib emphasizes this point in her critique of Butler's theoretical paradigm. Rejecting the "strong" version of postmodernism advocated by Butler, Benhabib accepts the insights of postmodern
theory only in their "weak" formulations. That is, she maintains that we can accept the constructedness of the subject by "various social, linguistic, and discursive practices" while also holding on to "the desirability and theoretical necessity of articulating a more adequate, less deluded, and less mystified vision
of subjectivity."[9] We can safeguard the important democratic capacities of the
humanist actor, "like self-reflexivity, the capacity for acting on principles, rational accountability for one's actions and the ability to project a life-plan into
the future," while also "taking account of the radical situatedness of the subject."[10] We can accept that "a subjectivity . . . not . . . structured by language,
by narrative and by the symbolic structures of narrative available in a culture

is unthinkable" while also maintaining that "we are not merely extensions of our histories," that "vis-à-vis our own stories we are in the position of author and character at once." We can recognize that "the situated and gendered subject is heteronomously determined" yet "still strives toward autonomy."[11] In short, not all notions of performativity require the complete renunciation of the humanist paradigm and its rich possibilities for agency. However, the concept of performativity as developed by Butler forecloses the possibility of radically reconstructing gender, because it denies the possibility of humanist subjects capable of rewriting their own scripts.

Despite these constraints, however, Butler strives to maintain some sense of agency. She stresses that the command to repeat may not always yield the type of "behavioral conformity" expected: "The call by the law that seeks to produce a lawful subject produces a set of consequences that exceed and confound what appears to be the disciplining intention motivating the law." Consequently, some possibilities for subversion exist even within Butler's matrix of constraints, including "the parodic inhabiting of conformity, . . . a repetition of the law into hyperbole, [and] a rearticulation of the law against the authority of the one who delivers it." Nevertheless, serious limits exist on the amount of agency possible within her framework: "This 'I' that is produced through the accumulation and convergence" of dominant discourses "*cannot extract itself from the historicity of that chain or raise itself up and confront that chain* as if it were an object opposed to me, which is not me, but only what others have made of me."[12] So while Butler's postmodern framework allows some possibility for change, it precludes the strong sense of agency characteristic of the humanist paradigm.

Furthermore, Butler's deconstructionist move also destroys the normative basis upon which to argue for the justness of making such changes. That is, while Butler advocates a politics of subversion, which she wants to put in the service of contesting those norms that she considers "cruel," without a normative theory of social justice and human dignity, she cannot provide a theoretical basis for distinguishing between subversive acts that further liberty, equality, and the rule of law and those that undermine those fragile ideals. To be sure, no one questions Butler's allegiance to politically progressive politics; however, as Martha Nussbaum rightly points out, "Butler cannot explain in any purely structural or procedural way why the subversion of gender norms is a social good while the subversion of justice norms is a social bad." Expanding on this line of argument, Nussbaum notes:

> There is a void, then, at the heart of Butler's notion of politics. This void can look liberating, because the reader fills it implicitly with a normative theory of human equality or dignity. But let there be no mistake: for Butler, as for Foucault, subversion is subversion, and it can in principle go in any direction. Indeed, Butler's naively empty politics is especially dangerous for the very causes she holds dear.

For every friend of Butler, eager to engage in subversive performances that pro-
claim the repressiveness of heterosexual gender norms, there are dozens of others
who would like to engage in subversive performances that flout the norms of tax
compliance, of nondiscrimination, of decent treatment of one's fellow students. To
such people we should say, you cannot simply resist as you please, for there are
norms of fairness, decency, and dignity that entail that this is bad behavior. But
then we have to articulate those norms—and this is what Butler refuses to do.[13]

In this passage, Nussbaum cogently articulates the major political problem with
a full-scale acceptance of the strong version of postmodernism. That is to say,
postmodernism does a good job of exposing the constructedness of the subject
by political, historical, and ideological forces beyond its control. However, the
postmodern rejection of humanist ideals as simply another mask for power actually eliminates the moral basis for opposing those particular forces that we
consider unjust.

Heeding Benhabib's and Nussbaum's critiques, *Citizen-Soldiers and Manly
Warriors* takes up the concept of performativity but places it within a normative
framework of social justice provided by the democratic tradition of civic republicanism. That is, I want to affirm the humanist ideals of civic republicanism—
such as the freedom, equality, and autonomy of the choosing subject—while simultaneously exposing the shortcomings of actually existing practices through
the limited deployment of some useful postmodernist conceptual tools. While I
am interested in subverting cultural constructions of gender that have functioned
to limit the participation of "women" in the public sphere, I am not interested in
subverting the entire tradition of civic republicanism, because it has historically
been exclusionary. Instead, I am interested in reconstructing republican citizenship so that *all* individuals, regardless of gender, can be included as republican
citizens. Placing performativity within the context of civic republicanism—with
its commitment to the universalizable principles of liberty, equality, camaraderie, the rule of law, civic virtue, and participatory citizenship—allows us to
differentiate forms of subversion that advance democracy—like the participation of "women" in historically "masculine" civic practices constitutive of citizenship—from those that undermine it—like the co-optation of "identity politics" by angry white men who oppose the extension of democracy to previously
excluded groups, as discussed in chapter 5.

This book uses a performativity lens to reread the normative tradition of
the Citizen-Soldier in order to raise the possibility of reconfiguring republican citizenship so that it meets the standards of true universality. That is to
say, if masculine citizen-soldiers are traditionally constituted through engagement in civic and martial practices, and feminine subjects are traditionally constituted through the exclusion from these same practices, then the engagement of "women" in the practices constitutive of masculine republican
citizenship should allow them to become citizen-soldiers on the same basis
as men. That is, the *subversive transgender performances* of "women" acting
out "male" scripts could work to highlight the artificiality of normative con-

structions of gender and consequently undermine the sexism such constructions generate.

Politically, this means that not only is there no reason for "women" not to participate in civic and martial practices culturally deemed "masculine," but in fact
such participation could radically undermine the traditional dichotomous construction of gender that has prevented women from achieving full equality. In
other words, the *subversive transgender performance* of "women" engaging in
the behaviors constitutive of citizen-soldiers should work to undermine the idea
that "men" and "women" must be restricted by the cultural imperatives of "masculinity" and "femininity," respectively. Moving beyond restrictive gender
norms will allow all individuals the freedom to live as they choose. The movement beyond the traditional dichotomous constructions of gender clears the way
for a proliferation of gender, which not only will undermine sexism but could also
allow full civic subjectivity for "women." That is, it could open up the possibility for "women" to become republican citizens on an equal basis with "men."

In beginning our exploration of the possibility for "women" to occupy the category of the Citizen-Soldier, we must first note that military service no longer at-
taches to citizenship within the American context. As I argued in chapter 4, by the
time of the Selective Draft Act of 1917, participatory citizenship had been dis-
connected from military service and had in fact been seriously undermined by the
rise of liberal individualism.[14] In other words, though all male citizens were required to serve in the military, this service was not coupled with the possibility of
substantive participation in self-government—that is, with republican citizenship.
Moreover, after the Vietnam War, mandatory service in the military was completely eliminated as a requirement of citizenship. We now have an all-volunteer
force of professional soldiers officially subordinate to civilian elected officials.
Yet despite these changes, engagement in martial practices within the U.S. military still produces *armed masculinity;* it is just not explicitly linked to citizenship.

Nevertheless, the Citizen-Soldier tradition remains our historical legacy and
so still forms the overall democratic context within which America imagines
its civil-military relations being situated. In other words, despite the historic di-
vorce of the Soldier from the Citizen documented in chapter 4, the Citizen-Soldier ideal still exists in American political culture. This partially explains why
many Americans on the Right fixate on the fact that President Clinton did not
serve in Vietnam.[15] And it also helps explain why many women, gay men, and
lesbians view military service as central to their acceptance as full citizens.
Building on Cynthia Enloe's contention that *armed masculinity* must be understood in its cultural particularity, I would like to suggest that American
armed masculinity must be understood as developing within the context of our
particular cultural version of the Citizen-Soldier ideal, replete with its inter-
woven set of virtues and vices.[16] Thus, we must approach the question of how
"women" can become citizen-soldiers by examining the contemporary
practices constitutive of *armed masculinity* within this broader context of civil-
military relations in America—the Citizen-Soldier tradition.

Feminists Theorize the Military

Most discussions of women and the military do not consider the Citizen-Soldier tradition. And most discussions of women and the military, even among feminist theorists, do not understand gender identity as performatively constituted. Instead, most theorists consider gender to be a core identity that pre-exists a person's relationship to the military. There are two basic versions of this idea. The first approach assumes that there are entities called *men* and *women* who then relate to the military in a variety of ways. Judith Hicks Stiehm's award-winning book *Arms and the Enlisted Woman* forms an excellent example of this first approach. The book studies "America's most unknown soldiers—enlisted women in the Army, Air Force, Navy, and Marines" in order to represent their experiences and make policy recommendations about how to make the military more equitable for women.[17] By assuming the existence of *men* and *women,* Stiehm does not examine the ways in which the military actually *produces* "armed masculinity"—how it makes men—or how women's exclusion from military practices contributes to the production of femininity.

In contrast, the *gender difference approach* is more theoretically sophisticated. It assumes that there are social constructs called "masculinity" and "femininity" that directly affect and are affected by the military and militarism. One version of this approach begins with the assumption of differential gender identities and explores the ways in which culturally constructed configurations of masculinity have shaped the military. Betty Reardon's classic book, *Sexism and the War System,* represents this school of thought. Building on the work of Carol Gilligan, Reardon posits the existence of masculine and feminine values that flow from men's and women's core gender identities.[18] Connecting masculine values to militarism, she argues that "the structures of violence that constitute the war system are . . . influenced by the attributes we use to guide the development of masculine identity and by masculine modes of public decision making."[19] In short, masculinity produces militarism. Consequently, "the feminine values, which nurture life and acknowledge the need for transcending competition and violence, are needed to guide policy formation to avoid or abolish war."[20] Understanding masculinity and femininity to be core identities rather than performative constructions, Reardon does not believe that women will lose their feminine values if they begin to engage in martial practices, such as making military policy. While Reardon provides a lot of important insights into militarism and its connection to sexism, she presents a static view of gender identity that reifies the traditional association of men with war and women with peace, thus playing right into sexism and militarism. She does not explore the ways in which martial practices actually *create* "armed masculinity" and how exclusion from these practices produces what we might term *pacific femininity.*

Cynthia Enloe exemplifies the second version of the *gender difference approach.*[21] Presenting a much more nuanced understanding of gender, Enloe argues that militarization relies on "varieties of masculinity and femininity," not just on one monolithic version of each. She argues that the military uses ideological ideals of masculinity to manipulate men and women into serving the needs of the military.[22] "Ignore gender—the social constructions of 'femininity' and 'masculinity' and the relations between them," she argues, "and it becomes impossible adequately to explain how military forces have managed to capture and control so much of society's imagination and resources."[23] Coming close to a performativity theory of gender, Enloe argues that "militarization is a gendered process . . . that won't 'work' unless men will accept certain norms of masculinity and women will abide by certain strictures of femininity."[24] That is to say, "militaries need women—but they need women to behave *as the gender 'women'*."[25] Here Enloe seems to suggest that gender is constructed and maintained through engagement in particular sets of gendered practices.

However, while Enloe presents a sophisticated, nuanced discussion of the multiple ways in which culturally constructed masculinities and femininities serve the needs of militarism, she does not recognize the possibility of *subversive transgender performances* and their potential to undermine traditional conceptions of gender. For example, in analyzing the role of the female soldier, Enloe does not consider the possibility that this type of counterhegemonic behavior could work to undermine the sex/gender system. For Enloe, the female soldier does not at all trouble our conflation of soldiering with masculinity with biological males. Instead, the female soldier simply exemplifies the "militarization of femininity."[26] Women will always be feminine, she implies, even if they repeatedly act out masculine scripts; they will just be feminine in different ways. Thus, Enloe theorizes a totalizing system that successfully co-opts every possibility of subversion.

Performativity theory adds three elements to these feminist approaches. First, it rejects the inevitability of a dyadic gender system and so holds out the possibility of eventually moving beyond gender—rendering gender irrelevant for civic purposes. Second, it replaces the idea of a core gender identity with the idea of gender as a process, as something that can never be finally achieved; one must constantly perform one's gender. Finally, performativity theory, as I have reinterpreted it, allows for the possibility of reconstructing gender through participation in *subversive transgender performances*. The spectacle of "men" acting "like women" and "women" acting "like men" should highlight the artificiality of supposedly natural manifestations of gender and, in this way, undermine the foundations of the sex/gender system and the sexism it generates. Thus, performativity theory challenges versions of feminism that begin with the assumption of sexual difference and that simply hope to valorize women's "different voice."

Jean Bethke Elshtain's work on women and the military hints at but does not develop the possibility of *subversive transgender performances*. Unlike Enloe and others who explore the global process of militarization in all its cultural particularities, Elshtain focuses on gender and military service within the

Western tradition of political thought and specifically within her reading of "armed civic virtue"—of the Citizen-Soldier tradition.[27] "We in the West," Elshtain argues, "are the heirs of a tradition that assumes an affinity between women and peace, between men and war, a tradition that consists of culturally constructed and transmitted myths and memories."[28] Indeed, the cultural constructs of *armed masculinity* and *pacific femininity*—or the Just Warrior and the Beautiful Soul as she calls them—are deeply rooted in our Western tradition and so are not easily dislodged.

Approaching a performativity theory of gender identity, Elshtain argues that the normative masculine ideal of the Just Warrior and feminine ideal of the Beautiful Soul require actual male and female individuals to behave in gender-appropriate ways. The gendered cultural ideals inherent in the Western tradition, she explains, require that "in time of war, real men and women . . . take on, in cultural memory and narrative, the personas of Just Warriors and Beautiful Souls." The ideals of the Just Warrior—"man construed as violent, whether eagerly and inevitably or reluctantly and tragically"—and the Beautiful Soul—"woman [construed] as nonviolent, offering succor and compassion"—that Elshtain articulates operate like our Citizen-Soldier ideal: Each prescribes a set of practices the participation in which constitutes biological males and females as masculine and feminine—as "men" and "women." In Elshtain's words, these ideals "do not denote what men and women *really* are in time of war, but *function instead to recreate and secure* women's location as noncombatants and men's as warriors."[29] The Just Warrior and the Beautiful Soul, like the Citizen-Soldier, are normative ideals that require biological males and females to *act like* "men" and "women"—and consequently, to *become* men and women.

Although in some ways less nuanced than Enloe's, Elshtain's analysis of *armed masculinity* actually surpasses Enloe's to the extent that it leaves open the possibility of the kind of *subversive transgender performances* I have been advocating. Confessing her own transgender fantasies—as a child, Elshtain dreamed of "swashbuckling . . . danger . . . handsome heroes and beautiful women" but she "gets to be a hero in those dreams, a woman in disguise, the greatest swordsman of them all"—Elshtain recognizes that males and females do not always identify in the normatively prescribed ways.[30] In her words, "these paradigmatic linkages [between men and war, and women and peace] dangerously overshadow other voices, other stories: of pacific males; of bellicose women; . . . of martial fervor at odds—or so we choose to believe—with maternalism in women."[31] Since *armed masculinity* and *pacific femininity* are cultural constructs rather than natural facts, there is no guarantee that men will become Just Warriors and women Beautiful Souls. Moreover, Elshtain seems to suggest that the continuation of normatively correct gender identification is not a sure thing: "No conscious bargain was struck by our collective foremothers and fathers to ensure [the traditional] outcome. Rather, sedimented lore—stories of male war fighters and women home keepers and designated weepers over war's inevitable tragedies — have spilled over from one epoch to the next."[32] Our tradition of "sedimented lore" exerts a powerful influence on us. Nevertheless, because it is cultural rather than natural, it cannot ensure that biological males will always identify with the Just Warrior rather than with the Beautiful Soul and biological females with the Beautiful Soul and never with the Just Warrior.

That current-day military officials also recognize, whether consciously or unconsciously, the possibility of *subversive transgender performances* is evidenced by the anxiety expressed in official military policies aimed at maintaining gender difference artificially. Why else would the U.S. Marine Corps require female soldiers to "tweeze their eyebrows in a regulation arch"? Why else would the caption under an Army recruitment poster featuring "a pretty woman smiling out from under a camouflaged combat helmet" read *Some of the best soldiers wear lipstick?*[33] Why else would the Army make "a policy that defines the carrying of umbrellas as 'unmanly' "?[34] Why else would West Point, upon becoming coeducational, decide to eliminate "running events in which cadets carry weapons?"[35] Why else would it modify "regulations relating to uniforms, hair length and jewelry"?[36] And why else would Citadel officials, after repeatedly threatening to shave cadet Shannon Faulkner's head, suddenly reverse themselves as soon as the young woman said she didn't care?[37]

Hazing Rituals in the American Military

As we have seen in previous chapters, the Citizen-Soldier ideal entails four key democratic elements. First of all, the Citizen-Soldier ideal embodies a set of practices that produce the necessary foundation for republican self-rule: The civic and martial practices constitutive of the Citizen-Soldier also produce patriotism, fraternity, civic virtue, and a common civic identity, the essential prerequisites for government aimed at the common good. Second, the Citizen-Soldier represents the idea that the soldiers who serve in the military are also the citizens who control the military — in short, that the civil should control the military. Third, the reason citizen-soldiers serve in the military is to defend their ability to govern themselves for the common good through the rule of law. And finally, the Citizen-Soldier constitutes a normative ideal. It means more than the fact that citizens comprise the military. The Citizen-Soldier also embodies a commitment to civic republicanism, complete with all its ideals: liberty, equality, fraternity, the rule of law, the common good, civic virtue, and participatory citizenship. Thus, the Citizen-Soldier ideal represents the link between participatory citizenship and military service.

One of the ways in which military service supposedly facilitates republican citizenship is by creating a sense of fraternity and common identity among previously diverse individuals. This fraternity entails a strong degree of bonding among its members, a bonding that is constituted through the rejection of previous identities and the acquisition of a new, common identity: the Manly Warrior. The feelings of fraternity and identification with the group that occur during military training lay the foundation for the emergence of martial virtues, such as selflessness, courage, heroism, and patriotism—which includes the willingness to risk one's life for the good of one's community or nation-state. Through engagement together in martial practices, diverse individuals become a fraternity of patriotic manly warriors.

Although the *armed masculinity* created in this way traditionally got fused to republican citizenship, as stressed earlier this is no longer the case. Today, military practices still produce *armed masculinity,* but it is not linked to participatory citizenship—a form of citizenship that, in any event, has become less possible. Consequently, the remainder of this section will focus on the creation of *armed masculinity* by contemporary military institutions.

Contemporary military practices still emphasize the creation of a common identity among previously diverse individuals. While the new identity of the Manly Warrior *could* supplement and enrich the previous identities of each individual, traditional military academies strive instead to break down and annihilate the previous identities. Sanford M. Dornbusch explains how this process works: "Successfully completing the steps in an Academy career . . . requires that there be a *loss of identity in terms of pre-existing statuses.* . . . This complete isolation [of the new cadet] helps *to produce a unified group of swabs, rather than a heterogeneous collection of persons* of high and low status. . . . *The role of the cadet must supersede other roles* the individual has been accustomed to play."[38] The cadet does not add his new identity to his old. He does not become, for example, a cadet at the Virginia Military Institute (VMI) *in addition* to being John Doe, a white, middle-class, Southern Baptist boy from Charlottesville, Virginia. Instead, he *loses* his previous identity: He is no longer John Doe, a white, middle class, Southern Baptist boy from Charlottesville. He is now simply a cadet at VMI. The identity of *cadet*—of Manly Warrior—*replaces* other categories of identification; it does not merely supplement them— at least according to the traditional method outlined by Dornbusch.

This method of unifying the group aims to create a sense of fusion among members. Cadets are supposed to be one and the same. Because of this overemphasis on fusion and homogeneity, the appearance of heterogeneity becomes a problem. As Dornbusch notes, writing in 1955, "[I]t is clear that the existence of minority-group status on the part of some cadets would tend to break down this desired equality."[39] Indeed, whenever a group of people who appear to be different wants to be integrated into the military, the desired fusion and homogeneity are believed to be threatened. Nevertheless, although the military vehemently resisted the integration of African Americans into its ranks, arguing that their presence would impede the process of creating fraternity and a common group identity, the military has very successfully overcome this challenge.[40] Will the same happen with straight women, gay men, and lesbians?

Traditionally, violent hazing rituals have facilitated the creation of a common identity within military academies. Dornbusch cites the important role hazing plays in the development of solidarity among cadets: "As a consequence of undergoing this very unpleasant experience together, the swab class develops remarkable unity." That is why hazing, although "forbidden by the regulations . . . is a hallowed tradition of the [military] Academy."[41] In a recent case, the all-male Virginia Military Institute argued in court that women should not be admitted because their presence would interfere with traditional hazing rituals—which is what happened when women entered West Point.[42] Moreover, although it did not say so in court, the Citadel must agree that hazing plays an essential role in military education, since it is notorious for not enforcing its official antihazing prohibitions.[43] Hazing remains an integral part of military education because it facilitates the creation of fraternal bonding.

The Virginia Military Institute sees hazing as central to its "unique" method of producing "citizen-soldiers." VMI employs what it calls an "adversative method" that they say is "intended to break down individualism and to instill the uniform values espoused by the institution."[44] According to the case record for *United States v. Commonwealth of Virginia,* "Colonel N. Michael Bissell, the Commandant of Cadets at VMI, summarized the educational process at VMI as follows: 'I like to think VMI literally *dissects the young student* that comes in there, kind of *pulls him apart.*' "[45] Living in the barracks, a VMI cadet "is totally removed from his social background."[46] The Citadel uses a comparable method.[47] It also strives "to 'strip' each young recruit of his original identity and remold him"—into a citizen-soldier.[48] Any identity the cadet had prior to entering the military academy must not remain intact.

Central to VMI's "adversative method" is the famous "rat line." According to court records, "entering students at VMI are called 'rats' because the rat is 'probably the lowest animal on earth.' In general, the rats are treated miserably for the first seven months of college. . . . The rat line is sufficiently rigorous and stressful that those who complete it feel both a sense of accomplishment and a *bonding to their fellow sufferers and former tormentors.*"[49] The rat line consists of "indoctrination, minute regulation of individual behavior, frequent punishments, rigorous physical education, and military drills."[50] Unrelenting subjection to the rat line "strips away cadets' old values and behaviors." After that, the "class system" uses peer pressure to teach and reinforce "the values and behaviors that VMI exists to promote."[51] The Citadel uses a similar process. As one young Citadel "knob" told Susan Faludi during her visit there: "We're all suffering together. It's how we bond."[52] The methods used at these two all-male public military academies instill in cadets a sense of "loyalty to one's brother rats"[53] or "knobs"—a sense of fraternity that underwrites the development of martial virtues.

Young male cadets at VMI and the Citadel become "citizen-soldiers" through a process that teaches them not only fraternity but also equality among peers, selflessness, and even a type of civic virtue—the willingness to put the good of the group ahead of one's own particular interest. "The VMI experience promotes . . . the belief that you must subordinate your own personal desires and well-being to the good of the whole unit."[54] Cadets learn a sense of self-lessness by being subjected to "a total lack of privacy" and by being "never free from scrutiny." While at VMI "a cadet cannot go to the bathroom or go to take a shower without being observed by everyone in that quadrangle on all levels."[55] Likewise, at the Citadel, Faludi tells us, "the sharing of the stall-less showers and stall-less toilets is 'at the heart of the Citadel experience,' according to more than one cadet." As one young man put it, 'I know it sounds trivial, but all of us in one shower, it's like we're all the same.'[56] All these ordeals produce "graduates who are prepared to be citizen-soldiers."[57]

VMI believes that the presence of women would interfere with its ability to create "citizen-soldiers." Although recently overruled by the United States Supreme Court, the United States District Court originally agreed with VMI, stating: " 'The mission of the Virginia Military Institute [is] to *produce educated and honorable men,* prepared for the varied work of civil life, . . . and ready as *citizen-soldiers* to defend their country in time of national peril.' . . . Excluding women is substantially related to this mission."[58] Why?

One of the reasons women must be excluded is that military institutions often use misogynistic and homophobic methods in constructing the Manly Warrior.[59] The presence of women (and lesbians and gays) would make the use of these methods problematic. While not essential to the production of soldiers, I would argue, misogynistic and homophobic hazing often plays a central role in military education at both military academies and in boot camps. Many scholars have documented the ways in which *armed masculinity* is explicitly constructed in opposition to femininity and homosexuality. For example, citing vivid examples from oral histories, Richard Moser documents the role of misogyny and homophobia in the constitution of the Manly Warrior identity. At boot camp, "the military's central socializing experience," Moser tells us, "violence and denigration were used as the introduction to the world of the soldier. . . . Violence and intimidation were mixed with strident and persistent appeals to sexual identity." Moser quotes ex-marine Jess Jesperson: "Especially in the earlier stages of boot camp, when people are real confused and real disorganized, they always said, 'Girls—you cunts—pussies.' " Based on his interview data, Moser concludes that

in boot camp sexist and homophobic appeals were used to train and discipline soldiers. In the exclusive all-male environment of boot camp, women were used as a negative example and positioned as the common "other." . . . Machismo, misogyny, and homophobia were employed both as exemplary ideals and as weapons to destroy competing forms of masculinity. . . . Both gay and feminine sexuality were used as threats and negative examples.[60]

Having women or gay men physically present makes problematic the process of turning them into the denigrated "other" and/or results in sexual harassment, rape, and other forms of violence.

Boot camps use methods similar to those of the military academies, as described by Dornbusch, Faludi, and VMI/Citadel students and officials. As Jesperson tells Moser: "It started when you went into boot camp. . . . There's a psychological terror . . . and physical torture. First, they dehumanize you, totally *take away your identity,* and then *remake you . . .* into what they want—*just* a fighter."[61] Boot camp, VMI, and the Citadel share a similar method of producing *armed masculinity.* Interestingly, VMI claims its "rat line" is even "more dramatic and more stressful than Army boot camp or Army basic training. . . . It is comparable to Marine Corps boot camp in terms of both the physical rigor and mental stress of the experience."[62] Thus, we can surmise that if boot camps routinely use misogyny and homophobia in constructing *armed masculinity,* then so does VMI, although no references were made to such tactics during its court cases.[63]

The Citadel, on the other hand, is notorious for its misogynistic and homophobic hazing. Faludi compiles plenty of evidence in her article. For example: " 'They called you a "pussy" all the time,' a young man who attended the Citadel in 1991 recalled. 'Or a "fucking little girl." ' " Faludi continues: "It started the very first day, when the upper-classmen stood around and taunted, 'Oh, you going to get your little girlie locks cut off?' " According to the former cadet's report, "virtually every taunt equated him with a woman: whenever he showed fear, they would say, 'You look like you're having an abortion,' or 'Are you menstruating?' "[64] These young "citizen-soldiers" forge their masculine identities in direct opposition to a denigrated femininity.

Using misogynistic and homophobic methods to construct fraternal bonds among soldiers creates a particularly unstable masculine identity predicated on the denigration of femininity and homoeroticism. While masculinity always forms in opposition to femininity, the type of *armed masculinity* created through the use of misogynistic and homophobic hazing defines itself as *hostile* toward what it perceives as the homoerotic and the feminine. This type of *armed masculinity* not only yields a hatred and fear of (all) women and gay men, but also requires that soldiers strongly repress the "feminine" parts of themselves as well as any homoerotic desire they might feel. Consequently, *armed masculinity* is a particularly precarious form of masculinity that always threatens to dissolve. Because *armed masculinity* can never be finally secured, the Manly Warrior must constantly engage in the practices constitutive of *armed masculinity:* He must constantly reestablish his masculinity by expressing his opposition to femininity and homoeroticism in himself and others.[65]

The anger, hostility, and aggressiveness produced in the process of constituting *armed masculinity* get channeled into a desire for combat against the enemy. As Moser found in his research, "machismo was employed as a gateway to other forms of domination and dehumanization. Once a sense of male superiority was achieved, then other forms of dominating behavior were introduced. . . . Dominance and manhood were equated with combat and opposition to the 'other.' This other was usually a 'pussy' and then a 'gook,' sometimes a 'commie,' but always a potential victim."[66] *Armed masculinity* serves militarism well.

Although the use of misogynistic and homophobic practices hurts and degrades people in the process of creating Manly Warriors out of individual civilians, the process is not completely devoid of pleasure—which of course is one of the reasons for its long-standing existence. In the first place, becoming part of something larger than oneself can be very pleasurable, even if the process involves pain. Secondly, many people enjoy dominating others, and those subjected to violent rituals will presumably get their chance to be on top in the future. And finally, some military men actually enjoy being subjected to painful and degrading rituals. For example, several of the sailors interviewed by Stephen Zeeland in *Sailors and Sexual Identity* reportedly enjoyed the "crossing the line ceremony" and similar rituals, even when they were the recipients of what many would consider unpleasant treatment. One man explains his experience as follows:

> I thought I was going to hate [the crossing the line ceremony], but I had fun. Some people quit, the people who couldn't take it anymore, men who were crying . . . it's really degrading. I had food all over me. I had Crisco oil poured on my face. I couldn't breathe, I got all this stuff in my lungs and in my eyes.[. . .] I liked it so much. . . . I just had a blast. Everyone thought I was going to break down. Everyone. . . . Some people were getting hit; the people who weren't liked very much were getting hit a lot. It got really disgusting. People were peeing in jars and saving it for weeks and pouring it on top of us. . . . I was on all fours, like a dog, and someone would be behind me actually hitting me with their dick like they were having sex with me. . . . I was laughing.[67]

A variety of pleasures exist in military discipline and violence, a reality we should not overlook if we want to understand our topic fully.

The intense nature of the fraternal bonding experience required for the creation of soldiers necessarily entails a certain amount of homoeroticism, even if unrecognized or explicitly denied by participants and others. While I am certainly *not* arguing that all soldiers are repressed homosexuals—a claim that would require an appeal to some sort of essential sexuality—I *am* suggesting that a repressed homoerotic tension inheres in many traditional military practices, and that this repressed sexual tension helps create both the bonding and the combativeness necessary for war-making. As Wilhelm Reich theorizes, "the goal of sexual suppression is that of producing an individual who is adjusted to the authoritarian order and who will submit to it in spite of all the misery and degradation. The suppression of natural sexual gratification leads to various kinds of substitute gratifications. Natural aggression becomes brutal sadism, which then is an essential mass-psychological factor in imperialistic wars."[68] The suppression and denigration of homoerotic desire and of any "feminine" aspects of the male's psyche plays a key role in the process of constructing the Manly Warrior.

The rituals involved in breaking down previous identities and reconstitut-ing the individuals as a fraternity of Manly Warriors are often *obviously* ho-moerotic. For example, Zeeland describes "Navy initiation rituals involving cross-dressing, spanking, simulated oral and anal sex, simulated ejaculation, nipple piercing, and anal penetration with objects or fingers," such as the fa-mous "crossing the line ceremony."[69] Similarly, many long-standing Citadel rituals are also clearly homoerotic, such as " 'Senior Rip-Off Day,' a spring rite in which three hundred seniors literally rip each other's clothes off, burn them in a bonfire, and hug and wrestle on the ground, . . . the birthday ritual, in which the birthday boy is stripped, tied to a chair, and covered with shav-ing cream, while his groin is coated in liquid shoe polish,"[70] and 'Bana-narama' during which "an unpeeled banana [is] produced—and shoved into a cadet's anus."[71] Homoeroticism plays a long-standing part in the military bonding process.

However, despite the homoeroticism of many traditional military practices, many of them are also clearly homophobic as well. Faludi provides an excel-lent example of this phenomenon: "Knobs told me that they were forced to run through the showers while the upperclassmen 'guards' knocked the soap out of their hands and, when the knobs leaned over to retrieve it, the upperclassmen would unzip their pants and yell, 'Don't pick it up, don't pick it up! We'll use you like we used those girls!' "[72] Submission to aggressive homoerotic hazing plays a central role in the process of *degrading* the individual, ripping apart his previous identity, and reconstituting him as part of a fraternity of Manly War-riors. Homoerotic desire must not be explicitly acknowledged. Instead, it must be repressed and channeled into aggression against the "other"—homosexuals, women, and "the enemy."

Although the hazing rituals just cited are clearly sexual in nature, Zeeland ar-gues that homoerotic desire exists even in military practices that are not ex-plicitly sexual. As he explains it,

> [H]omosexuality comes in many flavors, some known to the Joint Chiefs of Staff to be a natural part of military life. A desire to be in close quarters with other mil-itary men in a tightly knit brotherhood might be homosexual. . . . An officer's love for his men might be homosexual. The intimate buddy relationships men form in barracks, aboard ship, and most especially in combat—often described as being a love greater than between man and woman—might be homosexual—whether or not penetration and ejaculation ever occur.[73]

Zeeland proffers that the Pentagon opposes the presence of gay men in the mil-itary because it wants "to protect homoerotic military rituals, homosocial lifestyles, and covert military male-male sex from the taint of sexual suspicion [and]. . . . to maintain the illusion that there is not homosexuality in the mili-tary. This is the function of 'Don't Ask, Don't Tell': for boys to play with boys—and not get called queers, and not get called girls."[74]

Needless to say, military effectiveness requires that soldiers develop a sense of solidarity, feelings of affection for each other, and a recognition of their need for interdependence. Soldiers who are to go into battle together must learn to trust and depend on one another, and military academies know that. At the Citadel cadets learn interdependence in a variety of ways, including the "Citadel shirt tuck," which requires cadets to help each other dress. Many of Faludi's interviews with Citadel students and officials contain the theme of "manly nurturance." Even within the context of rigorous training, cadets nurture each other. Men at the Citadel hug and kiss each other. In fact one Citadel professor refers to the relationship between cadets as "like a true marriage. There's an affectionate intimacy that you will find between cadets. With this security they can, without being defensive, project tenderness to each other."[75] Through this bonding process individuals become a fraternity of Manly Warriors fit for military service.

A certain degree of homoeroticism, whether consciously recognized or not, must play a role in the creation of an effective combat unit. As one of the naval lieutenants interviewed by Zeeland put it:

> One thing I learned from being a really well-respected officer on-board ship is that there is a part of good military leadership that is inherently homosexual in nature. And that is love for your fellow man. . . . I think the Spartans understood that. It's only in our twentieth-century conservative Judeo-Christian mindset that we find it so incompatible with military service when in actual practice today—we don't call it homosexuality, but I think every good leader feels something of that. A yearning for his men. Not that it's consecrated physically, but it's everything short of that, and the feelings are the same.[76]

Feelings of homoerotic desire will necessarily arise in the intimate interactions of the military. But, as the lieutenant implies, these feelings of desire must also be repressed to some extent so that soldiers can focus on their mission rather than on their personal pleasures. However, when the military chooses to use misogynistic and homophobic tactics as a way of repressing homoerotic feelings that may arise, it produces an *armed masculinity* that is very precariously constructed against, as well as being hostile toward, femininity and homoeroticism. In that kind of military, women and gays must be excluded—a situation that is unacceptable in a democratic country.

The presence of women and/or gays in the military functions to highlight the homoerotic aspects of necessary military practices. When biological women are present, traditional military education and training practices suddenly appear sexualized. "The adversative method which pits male against male . . . would not produce the same results when a male is set against a female," argues VMI.[77] "Cross-sexual confrontation and interaction introduce additional elements of stress and distraction which are not accommodated by VMI's methodology."[78] And they are right. The tenor of "the adversative method"

changes when women participate. Imagine if an upperclassman treated a female "rat" in the same aggressive, misogynistic, and homophobic way he treats
a male "rat." He would look a lot like a sexual harasser—at best. Now imagine
if an upperclasswoman treated a male "rat" in the traditional way. She would
look a lot like a dominatrix. Without the physical presence of biological
women, military men can pretend that sexual desire plays no role in their practices. As VMI argues, "at an all-male college, adolescent males benefit from
being able to focus exclusively on the work at hand, *without the intrusion of
any sexual tension.*"[79] In any event, some traditional military practices look significantly different when the group is sexually mixed.

Routine military practices begin to look like sexual harassment when women
are involved. Lieutenant Colonel T. Nugent Courvoisie, former assistant commandant at the Citadel, told his wife that women should not be admitted because if they are, "there's going to be sexual harassment." In response, his wife
aptly quipped, "Oh, honey, those cadets are harassing each other right now, all
the time." The lieutenant colonel fired right back: "That's different. That's standard operating procedure."[80] And it is.

Now, one might plausibly argue that despite their sexually harassing nature,
many of the traditional military educational strategies are necessary to create the
intense kind of bonding required in a combat situation. This is not my argument,
but it is plausible. However, other practices commonly used at the Citadel are
completely indefensible. For example, an English professor at the Citadel argues
that "if Shannon were in my class, I'd be fired by March for sexual harassment.
. . . I've dealt with young men all my life. I know how to play with them. I have
the freedom here to imply things I couldn't with women. I don't want to have to
watch what I say."[81] What does he mean? Well, this professor likes to chastise
his students for using the passive voice, by way of the following: "Never use the
passive voice—it leads to effeminacy and homosexuality. . . . Next time you use
the passive voice I'm going to make you lift up your limp wrist."[82] This pedagogical method clearly serves no necessary military function.

Although the process the military academies use to produce masculine citizen-soldiers requires the repression and exclusion of femininity, it also needs
femininity. Femininity must exist so that it can be excluded and repressed. Because no biological women can be present, however, men must take turns
standing in for the absent women. For example, one of the sailors interviewed
by Zeeland says that while out at sea, he played the "sea bitch": "I was called
the sea bitch. That's just someone who—I think it's just a way for straight men
who have gay tendencies to let some of their frustration out. Because if they're
saying stuff toward me, it's nothing serious, 'cause I'm the sea bitch, right?
And if there was no sea bitch then they couldn't say it. . . . People would come
up to me and say, 'I'm gonna fuck you tonight. I'll pay a visit to your rack.' "[83]
Manly Warriors can establish their aggressive domination over women by
menacing the "sea bitch." A Citadel alumnus told Faludi that the traditional

indoctrination processes used at the Citadel entail "a submerged gender battle, a bitter but definitely fixed contest between the sexes, concealed from view by the fact that men played both parts. The beaten knobs were the women, 'stripped' and humiliated, and the predatory upperclassmen were the men, who bullied and pillaged. . . . [They cast] the male knobs in all the subservient feminine roles."[84]

The process productive of *armed masculinity,* as we have been describing it, gives rise to both the valorization of the myth of Woman and the vilification of actual women. Both Faludi and Moser discovered this dual view of women in their interviews with military men. For example, Moser found that "the sexual strategy of boot camp portrayed conventional feminine stereotypes as the polar opposite of the fighter-spirit" and that these stereotypes remain "abstract ideals."[85] However, as for actual women: "[G]irlfriends, wives, and mothers were commonly presented as sexually perverse. Woman soldiers too were stereotyped as 'whores or lesbians.' "[86] Likewise, Faludi learned that "the acknowledged policy [of the Citadel] is that women are to be kept at a distance so they can be 'respected' as ladies." As senior regimental commander, Norman Doucet explains, "the absence of women makes us understand them better." However, as Faludi documents, "women at less of a remove fare less well." Citing examples of disrespectfulness, sexual harassment, threats, and overt violence, Faludi concludes that "the Citadel men's approach to women seems to toggle between extremes of gentility and fury."[87] And, of course, the wide array of recent sex scandals that have plagued all branches of the American military should be understood as quintessential examples of what *armed masculinity* yields.

Reconstituting the Citizen-Soldier

The kind of *armed masculinity* we have been discussing presents a danger to women, gay men, and lesbians as well as to democratic citizenship. Clearly, we do not want to reinvigorate the Citizen-Soldier tradition in this country if it means reattaching a destructive, misogynistic, and homophobic *armed masculinity* onto citizenship. Moreover, the historically "masculine" category of the Manly Warrior cannot simply be expanded to include "women" but otherwise remain unaltered. The *armed masculinity* of contemporary soldiers is a precarious cultural construct constituted in hostile opposition to "femininity," whether located in "women" or within "men" themselves. Simply inserting "women" into a misogynistic warrior culture does not eliminate the conflation of soldiering with "masculinity," but rather produces sexual harassment and rape, as evidenced by the broad array of recent scandals within the American military. Because of traditional dichotomous constructions of gender, female individuals are viewed not as "soldiers" but as "women."

Nevertheless, there are two major reasons why we should not reject the Citizen-Soldier tradition in toto. First of all, it remains the only political theoretical discourse that recognizes the military as a central problem for democratic society. Hierarchical institutions of coercion, such as the police and the military, always pose a problem for a democratic society in which freedom and equality are fundamental values. Nevertheless, democratic societies require the existence of such institutions to protect themselves from those who would undermine the fragile ideals of liberty, equality, and the rule of law. Or to put it in terms of the tradition itself, a democracy must vigilantly guard itself against the threats posed by internal and external enemies. Because the Citizen-Soldier tradition directly addresses the potential contradiction between civic and martial imperatives, remembering this tradition should add a new dimension to current debates about the military, whereas overlooking this tradition allows us to refrain from taking the military seriously as a part of democratic society. Consequently, this tradition provides us with a set of democratic ideals with which we can strive to reform the military and purge it of misogyny and homophobia—neither of which is essential to military effectiveness.

Second, to reject the Citizen-Soldier tradition in its entirety would be to give up on the American tradition that anchors our calls for a more participatory form of citizenship—civic republicanism. The ideal of substantive popular sovereignty, as we saw in chapters 3 and 4, comes directly out of the civic republican tradition, which has at its center the Citizen-Soldier ideal. Because the grand foundationalist fictions of modernity have been challenged by postmodern thinkers, it has become harder for us to anchor normative claims in unshakable "foundations." For this reason, it is useful to work within an already existing tradition. Despite its many risks, the Citizen-Soldier tradition contains valuable democratic elements whose revival could greatly improve citizenship and democracy in America.

I want to rework the Citizen-Soldier tradition of civic republicanism, because it presents us with a tradition of participatory citizenship and a commitment to universalizable principles. Right now in America, the idea that we should have government for the common good has come under sustained attack. Our society often lacks the essential prerequisites for government aimed at the common good—patriotism, camaraderie, and civic virtue—because we rarely engage together in civic practices. Within the historic tradition of civic republicanism, diverse individuals—not diverse by today's standards, but each self-interested and unique in his own way—became citizens as they engaged together in civic practices. And while multicultural America presents more of a challenge, I believe it would be productive to consider the ways in which engagement together in civic practices today might constitute our diverse peoples as American citizens in a substantive, participatory republican sense.

Although the Citizen-Soldier tradition contains both virtues and vices, as I have amply demonstrated, I believe we can reconfigure the tradition in order to benefit from its virtues while simultaneously minimizing its vices. This re-working entails five key elements. First, we need to rid the military of misogynistic and homophobic hazing. The production of effective soldiers does not require such tactics. Camaraderie—a nongendered version of fraternity—can be constructed without falling back on simple yet antidemocratic and ultimately destructive methods. For example, many women's sports teams achieve the camaraderie and the common sense of purpose necessary for effective functioning without resorting to misogyny and homophobia.

Moreover, as we have seen, within the Citizen-Soldier tradition the military exists in order to defend democratic civil society from those who would destroy it. Yet despite this mission, many of the practices currently used in the production of manly warriors actually threaten to destroy democracy by fueling the antidemocratic forces of misogyny and homophobia. Thus, in evaluating the military, we must ask what effect current military practices have on democratic social relations. Clearly, it does not make sense to allow the military of a democratic country to sow the antidemocratic seeds of hatred among groups of citizens. The fundamental purpose for which a democratic society maintains a military is to protect itself from its enemies, not only external enemies but also those who would erode democracy from within—such as the misogynists and homophobes who oppose equality for women, gay men, and lesbians.

And finally, in a country founded on a commitment to universalizable principles, we cannot legitimately bar certain groups of citizens from participating fully in defense of their homeland and way of life. VMI and the Citadel are right that the admission of women to their institutions would destroy their long-standing traditions. But that is a good thing. Though a single-sex, publicly funded military education may or may not be beneficial for those it serves, such exclusionary institutions are simply unacceptable in a democracy.

One of the major risks of the Citizen-Soldier tradition, as we have seen both in this chapter and in our discussion of Rousseau, is an overemphasis on fusion, homogeneity, and the construction of a totalizing common identity that replaces all more particularistic identities. One easy way men bond with each other and forge a unitary and totalizing identity is through the exclusion of women. We can never completely eliminate the risks posed by the vicious side of the Citizen-Soldier tradition, but we do not have to engage in practices meant to encourage these vices—which is exactly what misogynistic and homophobic hazing practices do: They facilitate the creation of the worst vices of the Citizen-Soldier tradition—fusion, homogeneity, and a totalizing identity.

The second element necessary to the reconstitution of the Citizen-Soldier tradition involves a shift in emphasis from *military* service to military *service*. Traditionally, military service played a key role in the constitution of republican citizenship, because it was military service that, despite its sometimes deadly purposes, instilled in individuals the virtues necessary for self-government aimed at the common good—selflessness, courage, camaraderie, patriotism, and civic virtue. In arguing for the reconstitution of the Citizen-Soldier tradition, one of the changes I call for involves a shift in emphasis from *military* service to military *service*. On this point I join other democratic theorists, who advocate a modern-day transformation of the Citizen-Soldier tradition into a program of national or civic service. For example, Benjamin R. Barber argues that "universal citizen service could offer many of the undisputed virtues of military service: fellowship and camaraderie, common activity, teamwork, service for and with others, and a sense of community." But while citizen service would preserve the virtues of the Citizen-Soldier tradition, it would also minimize its corresponding vices: "In place of military hierarchy, it could offer equality; in place of obedience, cooperation; and in place of us/them conflict of the kind generated by parochial participation, a sense of mutuality and national interdependence."[88] Replacing military service with a broader vision of civic service would facilitate the inclusion of all Americans in the practices constitutive of republican citizenship and would thereby minimize the risks of fusion, homogeneity, and construction of a totalizing identity.

In addition, my rereading of civic republican theory through the lens of contemporary feminist theory strives to move away from the idea of citizenship as an *identity* and toward a reconceptualization of citizenship as a set of *civic practices*. Because identity always forms in opposition to what it excludes, emphasizing a common *identity* risks exacerbating the vicious side of the Citizen-Soldier tradition—its chauvinism, exclusion, and conformity. In addition, if this deep sense of civic *identity* is produced predominately through *military* service, this makes nationalistic military conquest more likely. In order to wage war, one must identify strongly as a member of a "people" or a "nation," and this type of deep identification most easily develops in opposition to an "enemy" on whom one wages war.

Shifting from an emphasis on *military* to an emphasis on *service* should help minimize the vices inherent in the Citizen-Soldier tradition and help change our definition of citizenship from a common identity to *participation in a set of civic practices*. Civic *service* does not require the same depth of identification as does *military* service. Participation in a wide variety of civic practices as one part of one's life produces a multidimensional, less totalizing form of identity. Engagement in such civic practices could constitute individuals as American citizens, but not as purely American and nothing else.[89] Moreover, situating military service within a broad array of civic practices should remind us that a democratic society has a military not only in order to defend its borders but also to defend its democratic principles, including equality and participatory citizenship.

The third element in my plan to reconstitute the Citizen-Soldier tradition requires instituting a program of civic education in all military training. Instead of military service being the primary form of civic education, civic education should become a primary part of military service. If individuals want to bear arms in defense of America, they need to know exactly what they are defending, exactly what America stands for. Although a lot of debate exists around this question, I join a whole host of democratic theorists who underline the centrality of universalizable principles, democratic republican ideals, and the concept of popular sovereignty to what America means.

The fourth element necessary to reconstituting the Citizen-Soldier tradition involves remembering the view of civil-military relations inherent in the ideal of the Citizen-Soldier. Traditionally, this ideal fused the militia onto the civic realm of republican self-rule. The Soldier who risked his life to defend the republic was also the Citizen who participated in forming laws for the common good. Both halves of the Citizen-Soldier ideal were equally important: Citizen-soldiers fought to defend their ability to govern themselves for the common good through the rule of law. In other words, the Citizen-Soldier ideal does not mean simply that citizens comprise the military. Normative rather than empirical, the Citizen-Soldier embodies a commitment to civic republicanism, complete with all its ideals: liberty, equality, camaraderie, the rule of law, the common good, civic virtue, and participatory citizenship.

While the realities of contemporary politics have necessitated a move away from a reliance on a system of citizens' militias and toward a professionalized, technologically trained military, we have not completely abandoned the American understanding of the Citizen-Soldier tradition: that the military should be subordinate to elected civilian officials, who should be accountable to the American citizenry. Within this context, I advocate a resuscitation of American citizenship to include more substantive participation in government. Whether or not we wage war or send troops to participate in UN peacekeeping projects should be decided through civic deliberation among citizens. This is the true spirit of the Citizen-Soldier tradition.[90]

My final suggestion for reconstituting the Citizen-Soldier tradition entails seizing the radical opportunities for "troubling" cultural constructions of gender inherent in a tradition that implicitly contains a performative understanding of civic and gender identity. In order to do this, we must, first, replace *armed masculinity* with a new form of *civic masculinity* and, second, make "gender trouble" by encouraging biological females to engage in the practices constitutive of *civic masculinity*. Simultaneously, we should encourage biological males to participate in the practices of a revitalized *robust femininity*. These types of *subversive transgender performances* should help realize the radical democratic potential for reconstructing the sex/gender system inherent in the civic republican tradition as I have reread it in this study.

That is to say, if masculinity is a cultural construct rather than a natural attribute of biological males, female individuals *theoretically* should be able to participate in these practices and become citizen-soldiers alongside men. The *subversive transgender performance* of "women" engaging in culturally "masculine" practices would highlight the artificiality of gender and thus undermine the idea that "masculinity" and "femininity" are natural attributes of biological males and females, respectively, the idea that underlies the sex/gender system and the sexism it generates. Instead of being restricted by sexist societal imperatives, individuals would be free to perform gender however they choose. In other words, a radical democratic moment exists in "women" acting "like men" (and vice versa).

However, this radical democratic possibility of "gender trouble" cannot be realized when the practices constitutive of Manly Warriors are inherently misogynistic and homophobic. Women, gay men, and lesbians cannot be included in an *armed masculinity* constituted through the denigration of femininity and homoeroticism. This forms yet another reason to eliminate misogynistic and homophobic hazing practices from the military and lift the combat exclusion: so that we can make "gender trouble" for the purpose of reconstructing citizenship so that it meets the standards of true universality.

Reimagining the Civic Public

In other words, instead of continuing to create *armed masculinity* and trying to reattach it to a resuscitated participatory citizenship, we should strive to create a new form of masculinity, called *civic masculinity,* that *all* people, male and female, can perform. Notice that I am maintaining the term "masculine" and calling for the participation of female individuals in "masculine" practices as a way of highlighting the destabilization of traditional gender norms that have historically prevented "women's" participation in republican citizenship. Eliminating the terms "masculine" and "feminine" altogether, though eventually the goal, at this point would truncate the radical democratic potential to make "gender trouble" in a tradition that entails a performative theory of civic and gender identity.

As I am imagining it, *civic masculinity* would entail the virtues, capacities, and pleasures traditionally associated with citizenship and the public sphere — where ideally all citizens would participate in the forming of political judgments within a context of universalizable republican principles, such as liberty, equality, and the rule of law. As a counterpart to this, I imagine a *robust femininity* that encompasses the virtues, capacities, and pleasures traditionally associated with the personal, domestic, and erotic spheres — such as nurturance, love, reproduction, spirituality, eroticism, playfulness, and the special attachments we reserve for particular people — all of which should be protected from unwanted political incursions. In other words, all people, both male and female, would become both *civicly masculine* and *robustly feminine,* for all would participate as citizens in political rule and in the personal realm with those with whom they choose to make their lives. This radical reconfiguration of the sex/gender system would undermine the idea that female individuals should not participate fully in civic practices as well as the idea that male individuals do not have to participate fully in domestic affairs.

I clearly want to maintain the modern concept of the public/private split to some extent, but I do not believe that we should have to renounce our particular affiliations as a prerequisite to participation in civic practices. We should not have to abstract from our particularities in order to attempt to achieve a nongendered, nonraced, nonclassed *civic identity*. Instead, we must enter the civic process from where we stand. It is only through actually engaging with others that can we participate in the construction of a shared vision. Again, I am suggesting that we shift away from the idea of citizenship as an identity and begin to reconceptualize it instead as participation in a set of *civic practices*.

In addition, though I want to maintain many of the ideals of modernity—such as, universalizable principles, rational laws, a neutral state, and a private sphere of personal autonomy free from government incursion—I also want to avoid a radical dichotomy between reason/emotion and public/private. That is to say, although we probably want to be governed by rational laws, we also need to allow some room for emotions when making reasonable political judgments. Some emotions should be considered in political discussions and given moral weight, such as feelings of sympathy, compassion, generosity, and mercy. Even potentially dangerous emotions, such as feelings of fear, anger, outrage, and resentment, should weigh into political discussions so that they can be tempered by reasoned deliberation. The emotions cannot and must not be completely discounted in politics. However, they must be interrogated via reasoned conversation, reflected upon, and evaluated within the parameters of a commitment to universalizable republican principles. Furthermore, I want to maintain a private sphere of personal autonomy without having to ban all so-called "private" or "personal" issues from the political conversation—issues concerning sexual harassment, rape, and incest, for example. These political problems must be deliberated upon, just like any other political issues.

One of the virtues of the civic republican tradition, as I have shown throughout this study, is that it gives the passions a constructive role to play in politics. In order to be able to govern for the common good, a person must become emotionally attached to his (or her) community. On this point, Iris Marion Young and Joan Landes misunderstand the civic republican tradition. In her discussion of the civic public, Young criticizes Rousseau and by extension the civic republican tradition for "instituting a moral division of labor between reason and sentiment, identifying masculinity with reason and femininity with sentiment and desire. . . . By assuming that reason stands opposed to desire, affectivity and the body, the civic public must exclude bodily and affective aspects of human existence." This dichotomy profoundly affects citizenship, she continues, because it results in the exclusion "from the public [of] those individuals and groups that do not fit the model of the rational citizen who can transcend body and sentiment."[91] Landes makes a similar argument in her analysis of the Rousseauian "public sphere."[92] In opposition to these faulty characterizations of the civic republican tradition, I have demonstrated, particularly in chapter 3, that passion, pleasure, and desire are not in fact purged from the civic public sphere. Instead, they play a key role in civic republicanism: They are channeled into the civic virtues of patriotism, fraternity, and concern for the common good; these feelings underwrite the willingness and ability of individuals to govern themselves in accordance with the common good.

I also disagree with Landes and Young on the issue of whether the civic republican concept of the "public sphere" or "civic public" are *inherently* masculine constructs. Landes argues that in its inception the "public sphere"—the sphere in which citizens engage in civic deliberation—was formed against and not just without women. While this is undoubtedly true historically, as I myself have argued in previous chapters, I disagree with Landes's conclusion: That the public sphere *requires* women's exclusion. In her words, a public sphere based on "reason was counterposed to femininity, *if* by the latter we mean (as contemporaries did) pleasure, play, eroticism, artifice, style, politesse, refined facades, and particularity."[93] Because the public sphere excludes "femininity," Landes argues, it can never include women; it is essentially and irretrievably masculine. Taking issue with those who argue that previously masculine categories can now be universalized, she states that "the claim that the 'modernity' inaugurated in part by the French Revolution has 'not yet' exhausted its liberatory potential is . . . suspect. In the historically oriented critique of the public sphere here projected, this claim can never be redeemed, for the women's movement cannot 'take possession' of a public sphere that has been *enduringly* reconstructed along masculinist lines."[94] And Young concurs.[95]

Both Landes and Young argue that the civic republican tradition *necessarily* excludes women—even in the late twentieth century—because its very categories are constructed against women and not just without them. But this seems to imply some sort of essentialist thinking about gender. That is, while both Landes and Young clearly understand the constructedness of gender—and demonstrate the ways in which gender is constructed—they seem to imply that biological males and females are inextricably bound to particular historical constructions of masculinity and femininity, respectively. For example, when Landes argues that the concept of the *public sphere* was constructed against femininity and by extension against women, and that therefore the public sphere can never encompass women, she implies that women are inextricably linked to a "femininity" defined as "pleasure, play, eroticism, artifice, style, politesse, refined facades, and particularity." While she convincingly argues that the eighteenth-century public sphere was in fact constructed against and not just without a culturally defined femininity, this does not mean, I would argue, that therefore biological females can never participate fully in the public sphere. To the contrary, the idea of social construction means that what is constructed can be reconstructed—a point both Landes and Young would surely have to concede on the basis of their own arguments about the constructedness of gender. Therefore, the socially constructed linkage of "woman" with a "femininity" defined as the private, the particular, and the passionate, and the socially constructed linkage of "man" with a "masculinity" defined as the public, the universal, and the reasonable, while undoubtedly strong in our culture, can, nevertheless, be broken. That is, if gender is culturally constructed, then "women" are not inextricably bound to any particular historically constructed articulation of "femininity"—no matter how entrenched.

The cultural meaning of "femininity" as "pleasure, play, eroticism, artifice, style, politesse, refined facades, and particularity" does not follow automatically from biological female sex. Instead, as I have been arguing, it is the engagement in feminine practices and the lack of engagement in masculine practices that actually produce "women," whose existence then confirms the "naturalness" of this cultural construction. Likewise, the engagement of biological males in civic and martial practices actually produces masculine citizen-soldiers. This understanding of the performative construction of categories of identity, such as "men," "women," and "citizen-soldier," opens up a space in which these constructions can be contested and transformed.

In a tradition that conflates masculinity and citizenship, the attempts of "women" to engage in civic practices constitutes transgressive behavior. It is my contention that the culturally constructed categories of "femininity" and "masculinity" can be contested and (at least partially) transformed through the transgressive behavior and performance of nonnormative gender roles—by *subversive transgender performances*. If "masculinity" is defined as what is appropriate to the political sphere—as Landes demonstrates it was at the dawn of modernity—and "femininity" is defined as what is excluded from the political sphere, this does not mean that biological females must necessarily be excluded from the political sphere. Instead, it means that we must sever the link between biological males and "masculinity" and that between biological females and "femininity" and redefine "masculinity" and "femininity" as sets of *practices* in which we all engage at particular times, rather than as embodied identities. In order for this to happen, *all* people need to engage in the practices of our new configurations of *civic masculinity* and *robust femininity*.

Again, the reason I want to hold on to the term "masculine" when referring to civic virtues, capacities, and pleasures is because doing so allows room for the *subversive transgender performances* that will undermine the sex/gender system and the sexism it generates. This means that biological females should be encouraged to participate in the democratically reformed military I have advocated. As an additional benefit, this should help the country get the best military personnel it can. As Elshtain argues in *Women and War,* allowing full female participation in the military by lifting all combat restrictions—and eliminating misogynistic and homophobic practices—is "one way to relocate male and female selves to provide for a freer play of individual and civic capacities."[96] One of the things Elshtain wants is to "free up identities, offering men and women the opportunity to share risks as citizens," and, if they want, "to take up nonviolence as a choice, not a given" based on their biology.[97] Like Elshtain, I want all individuals—"men" and "women"—to participate equally in civic practices. However, unlike Elshtain, I would not consider this vision of "women's" participation one of the "several ways to occupy the category *woman*."[98] Instead, I would consider it a way to blow apart the categories of *woman* and *man* and thus clear the way for the proliferation of genders and the participation of *all* people, as *citizens,* in the civic republican dream of liberty, equality, camaraderie, the rule of law, the common good, civic virtue, and participatory citizenship.

Notes

1. For women's struggle to bear arms, see Darline Levy and Harriet B. Applewhite, "Women and Militant Citizenship in Revolutionary Paris," in *Rebel Daughters: Women and the French Revolution,* ed. Sara E. Mezler and Leslie W. Rabine (New York: Oxford University Press, 1992); "Women, Radicalization and the Fall of the French Monarchy," in *Women and Politics in the Age of the Democratic Revolution,* ed. Harriet B. Applewhite and Darline Levy (Ann Arbor: University of Michigan Press, 1993); and Dominique Godineau, "Masculine and Feminine Political Practice during the French Revolution, 1793-Year III," in *Women and Politics in the Age of the Democratic Revolution.* For women who dressed as men to serve in the military, see Julie Wheelwright, *Amazons and Military Maids: Women Who Dressed as Men in Pursuit of Life, Liberty and Happiness* (London: Pandora Press, 1989). Also see Cynthia Enloe, *Does Khaki Become You? The Militarization of Women's Lives* (Boston: South End Press, 1983), 120.
2. Judith Butler, *Gender Trouble: Feminism and the Subversion of Identity* (New York: Routledge, 1990), 6.
3. Butler, *Gender Trouble,* 33.
4. Butler, *Gender Trouble,* 33–34, emphasis mine.
5. Judith Butler, *Bodies that Matter: On the Discursive Limits of "Sex"* (New York: Routledge, 1993), x.
6. As I define it, postmodernism includes the wide diversity of theories that contest the basic premises of political modernity. This would include Butler's work, although she personally rejects the term *postmodernism,* which she considers too monolithic, in favor of the term *poststructuralism.* See Judith Butler, "Contingent Foundations," in *Feminist Contentions: A Philosophical Exchange,* ed. Seyla Benhabib, Judith Butler, Drucilla Cornell, and Nancy Fraser (New York: Routledge, 1995), 35–39.
7. Butler, *Bodies that Matter,* 94–95.
8. Butler, *Bodies that Matter,* 95; *Gender Trouble,* 142
9. Seyla Benhabib, "Feminism and Postmodernism," in *Feminist Contentions: A Philosophical Exchange,* ed. Seyla Benhabib, Judith Butler, Drucilla Cornell, and Nancy Fraser (New York: Routledge, 1995), 20.
10. Benhabib, "Feminism and Postmodernism," 20.
11. Benhabib, "Feminism and Postmodernism," 20–21.
12. Butler, *Bodies that Matter,* 122, emphasis mine.

13. Martha Nussbaum, "The Professor of Parody: The Hip Defeatism of Judith Butler," *New Republic,* 22 February 1999, 43.

14. For a full discussion of this concept, see Sara M. Evans and Harry C. Boyte, *Free Spaces: The Sources of Democratic Change in America* (Chicago: University of Chicago Press, 1992).

15. For a discussion of the role candidates' military service plays during political campaigns, see Sheila Tobias, "Shifting Heroisms: The Uses of Military Service in Politics," in *Women, Militarism, and War: Essays in History, Politics, and Social Theory,* ed. Jean Bethke Elshtain and Sheila Tobias (New York: Rowman & Littlefield, 1990).

16. Cynthia Enloe, "Beyond 'Rambo': Women and the Varieties of Militarized Masculinity," in *Women and the Military System,* ed. Eva Isaksson (New York: St. Martin's Press, 1988).

17. Judith Hicks Stiehm, *Arms and the Enlisted Woman* (Philadelphia: Temple University Press, 1989).

18. For Carol Gilligan's classic work, see *In a Different Voice: Psychological Theory and Women's Development* (Cambridge: Harvard University Press, 1982).

19. Betty Reardon, *Sexism and the War System* (New York: Teachers College Press, 1985), 89.

20. Reardon, *Sexism and the War System,* 4.

21. For another example of this approach, see Jacklyn Cock, *Colonels and Cadres: War and Gender in South Africa* (Cape Town, South Africa: Oxford University Press, 1991).

22. Enloe, "Beyond 'Rambo'," 73–74.

23. Enloe, *Does Khaki Become You?* 212.

24. Cynthia Enloe, "Bananas, Bases, and Patriarchy," in *Women, Militarism, and War: Essays in History, Politics, and Social Theory,* ed. Jean Bethke Elshtain and Sheila Tobias (New York: Rowman & Littlefield), 202.

25. Enloe, *Does Khaki Become You?* 212.

26. Cynthia Enloe, *The Morning After: Sexual Politics at the End of the Cold War* (Berkeley: University of California Press, 1993), 206–7.

27. For my disagreement with Elshtain's reading of the Citizen-Soldier tradition, see chapter 3 of this work.

28. Jean Bethke Elshtain, *Women and War* (New York: Basic Books, 1987), 4.

29. Elshtain, *Women and War,* 4.

30. Elshtain, *Women and War,* 16.

31. Elshtain, *Women and War,* 4.

32. Elshtain, *Women and War,* 4.

33. Enloe, *Does Khaki Become You?* 119.

34. Enloe, *The Morning After,* 51.

35. *United States v. Commonwealth of Virginia,* 766 F.Supp. 1407 (W.D.Va 1991), 1439.

36. *United States v. Commonwealth of Virginia,* 766 F.Supp. 1441.

37. See "Women without Hair: Lost or Found?" *New York Times,* 7 August 1994; "Judge Allows Head Shaving of a Woman at the Citadel," *New York Times,* 1 August 1995; "Storming the Citadel," *USA Weekend,* 28–30 July 1995; "Judge Rules the Citadel May Shave Woman's Head," *New York Times,* 10 August 1995; and "A Woman Reports for Duty as a Cadet at the Citadel," *New York Times,* 13 August 1995.

38. Sanford M. Dornbusch, "The Military Academy as an Assimilating Institution," *Social Forces* 33 (1955): 317.

39. Dornbusch, "The Military Academy," 317.

40. See Charles C. Moskos and John Sibley Butler, *All that We Can Be: Black Leadership and Racial Integration the Army Way* (New York: Basic Books, 1996).

41. Moskos and Butler, *All that We Can Be,* 318.

42. *United States v. Commonwealth of Virginia,* 766 F.Supp. 1407 (W.D.Va 1991), 1441.

43. Susan Faludi, "The Naked Citadel," *New Yorker,* 5 September 1994, 62–81.

44. *United States v. Commonwealth of Virginia,* 44 F.3d 1229 (4th Cir. 1995), 1239.

45. *United States v. Commonwealth of Virginia,* 766 F.Supp. 1407 (W.D.Va 1991), 1421.

46. *United States v. Commonwealth of Virginia,* 766 F.Supp, 1424.

47. *Faulkner v. Jones,* 10 F.3d 226 (4th Cir. 1993), 229.

48. Faludi, "Naked Cadet," 64.

49. *United States v. Commonwealth of Virginia,* 766 F.Supp, 1422, emphasis mine.

50. *United States v. Commonwealth of Virginia,* 976 F.2d 890 (4th Cir. 1992), 1239.

51. *United States v. Commonwealth of Virginia,* 766 F.Supp. 1407 (W.D.Va 1991), 1423.

52. Faludi, "Naked Citadel," 64.

53. *United States v. Commonwealth of Virginia,* 766 F.Supp. 1407 (W.D.Va 1991), 1423.

54. *United States v. Commonwealth of Virginia,* 766 F.Supp, 1427.

55. *United States v. Commonwealth of Virginia,* 766 F.Supp, 1424.

56. Faludi, "Naked Citadel," 64.

57. *United States v. Commonwealth of Virginia,* 766 F.Supp. 1407 (W.D.Va 1991), 1427.

58. *United States v. Commonwealth of Virginia,* 766 F.Supp, 1425, 1413

59. See Richard R. Moser, *The New Winter Soldiers: GI and Veteran Dissent during the Vietnam Era* (New Brunswick, NJ: Rutgers University Press, 1996), chap. 2; Faludi, "Naked Citadel"; and Steven Zeeland, *Sailors and Sexuality: Crossing the Line Between 'Straight' and 'Gay' in the U.S. Navy* (New York: Haworth Press, 1995).

60. Moser, *New Winter Soldiers,* 26–27.

61. Moser, *New Winter Soldiers,* 29, emphasis mine.

62. *United States v. Commonwealth of Virginia,* 766 F.Supp. 1407 (W.D.Va 1991), 1422.

63. A young midshipman at the United States Naval Academy confirmed that misogyny and homophobia are still strong at the USNA. Personal conversation, 27 October 1995.

64. Faludi, "Naked Citadel," 70.

65. As Moser puts it, "the obsessive drive to create and maintain machismo drew upon an insatiable insecurity that may be momentarily slaked only by a display of domination against some threat," *New Winter Soldiers,* 28.

66. Moser, *New Winter Soldiers,* 29.

67. Zeeland, *Sailors and Sexual Identity,* 57–58.

68. Wilhelm Reich, *Mass Psychology of Fascism* (New York: Farrar, Straus and Giroux, 1970), 31.

69. Zeeland, *Sailors and Sexual Identity,* 5.

70. Faludi, "Naked Citadel," 79.

71. Faludi, "Naked Citadel," 80.

72. Faludi, "Naked Citadel," 80.

73. Zeeland, *Sailors and Sexual Identity,* 5.

74. Zeeland, *Sailors and Sexual Identity,* 5–6.

75. Faludi, "Naked Citadel," 79.

76. Zeeland, *Sailors and Sexual Identity,* 65.

77. *United States v. Commonwealth of Virginia,* 44 F.3d 1229 (4th Cir. 1995), 1233.

78. *United States v. Commonwealth of Virginia,* 976 F.2d 890 (4th Cir. 1992), 896.

79. *United States v. Commonwealth of Virginia,* 766 F.Supp. 1407 (W.D.Va 1991), 1435, emphasis mine.

80. Faludi, "Naked Citadel," 78.

81. Faludi, "Naked Citadel," 78.

82. Faludi, "Naked Citadel," 78–79.

83. Zeeland, *Sailors and Sexual Identity,* 50.

84. Faludi, "Naked Citadel," 70.

85. Moser, *New Winter Soldiers,* 27–28.

86. Moser, *New Winter Soldiers,* 28.

87. Faludi, "Naked Citadel," 70.

88. Benjamin R. Barber, *Strong Democracy: Participatory Politics for a New Age* (Berkeley: University of California Press, 1984), 302. For a realistic and detailed plan for the "reconstruction of the citizen-soldier," see Charles C. Moskos, *A Call to Civic Service: National Service for Country and Community* (New York: Free Press, 1988).

89. For an interesting discussion of this idea, see Michael Walzer, *What It Means to Be an American: Essays on the American Experience* (New York: Marsilio Publishers, 1996).

90. In fact that is why Moser makes the interesting if controversial argument that the Vietnam war resisters were the real citizen-soldiers of that era because they advocated citizen control over military decision making. They wanted to reattach participatory citizenship to military service.

91. Walzer, *What It Means to Be an American,* 66.

92. See Joan Landes, *Women and the Public Sphere in the Age of the French Revolution* (Ithaca: Cornell University Press, 1988).

93. See Landes, *Women and the Public Sphere,* 46, emphasis mine.

94. See Landes, *Women and the Public Sphere,* 202, emphasis mine.

95. Iris Marion Young, "Impartiality and the Civic Public," in *Feminism as Critique,* ed. Seyla Benhabib and Drucilla Cornell (Minneapolis: University of Minnesota Press, 1988), 66.

96. Elshtain, *Women and War,* 244.

97. Elshtain, *Women and War,* 257.

98. Elshtain, *Women and War,* 232.

Selected Bibliography

Arendt, Hannah. *On Revolution.* London: Penguin Books, 1963.

Avineri, Shlomo. *Hegel's Theory of the Modern State.* Cambridge: Cambridge University Press, 1972.

Bailyn, Bernard. *The Ideological Origins of the American Revolution.* Cambridge: Belknap Press of Harvard University Press, 1992.

Barber, Benjamin R. *Strong Democracy: Participatory Politics for a New Age.* Berkeley: University of California Press, 1984.

———. "Unscrambling the Founding Fathers." *New York Times Book Review,* 13 January 1985.

———. *A Place for Us: How to Make Society Civil and Democracy Strong.* New York: Hill and Wang, 1998.

Baron, Hans. "Machiavelli: Republican Citizen and Author of the *Prince.*" *English Historical Review* 76 (1961): 217–53.

———. *The Crisis of the Early Italian Renaissance,* 2d ed. Princeton: Princeton University Press, 1966.

Bayley, C. C. *War and Society in Renaissance Florence.* Toronto: University of Toronto Press, 1961.

Benhabib, Seyla, Judith Butler, Drucilla Cornell, and Nancy Fraser, eds. *Feminist Contentions: A Philosophical Exchange.* New York: Routledge, 1995.

Berlant, Lauren. *The Anatomy of National Fantasy: Hawthorne, Utopia, and Everyday Life.* Chicago: University of Chicago Press, 1991.

Blum, Carol. *Rousseau and the Republic of Virtue: The Language of Politics in the French Revolution.* Ithaca: Cornell University Press, 1986.

Bock, Gisela; Quentin Skinner; and Maurizio Viroli; eds. *Machiavelli and Republicanism.* Cambridge: Cambridge University Press, 1993.

Brown, Wendy. *Manhood and Politics: A Feminist Reading in Political Theory.* Totowa, NJ: Rowman & Littlefield, 1988.

Butler, Judith. *Gender Trouble: Feminism and the Subversion of Identity.* New York: Routledge, 1990.

———. *Bodies That Matter: On the Discursive Limits of "Sex."* New York: Routledge, 1993.

Cassirer, Ernst. *The Myth of the State.* New Haven: Yale University Press, 1946.

169

Chambers, John Whiteclay, II. *To Raise an Army: The Draft Comes to Modern America.* New York: Free Press, 1987.

————. *Tyranny of Change: America in the Progressive Era 1890–1920,* 2d ed. New York: St. Martin's Press, 1992.

Clark, Richard C. "Machiavelli: Bibliographic Spectrum." *Review of National Literatures* 1 (1970): 93–135.

Coates, James. *Armed and Dangerous: The Rise of the Survivalist Right.* New York: Hill and Wang, 1995.

Cochrane, Eric. "Machiavelli: 1940–1960." *Journal of Modern History* 33 (1961): 113–36.

Crackel, Theodore J. *Mr. Jefferson's Army: Political and Social Reform of the Military Establishment, 1801–1809.* New York: New York University Press, 1987.

Cress, Lawrence. *Citizens in Arms: The Army and the Militia in American Society to the War of 1812.* Chapel Hill: University of North Carolina Press, 1982.

Croce, Benedetto. *Politics and Morals.* Translated by Salvatore J. Castiglione. New York: Philosophical Library, 1945.

DePugh, Robert B. "Political Platform of the Patriotic Party." In *Extremism in America,* edited by Lyman Tower Sargent. New York: New York University Press, 1995.

Diamond, Sara. *Roads to Dominion: Right-Wing Movements and Political Power in the United States.* New York: Guilford Press, 1995.

Dornbusch, Sanford M. "The Military Academy as an Assimilating Institution." *Social Forces* 33 (1955): 316–21.

Elshtain, Jean Bethke. *Public Man, Private Woman: Women in Social and Political Thought.* Princeton: Princeton University Press, 1981.

————. *Women and War.* New York: Basic Books, 1987.

Enloe, Cynthia. *Does Khaki Become You? The Militarization of Women's Lives.* Boston: South End Press, 1983.

————. "Beyond 'Rambo': Women and the Varieties of Militarized Masculinity." In *Women and the Military System,* edited by Eva Isaksson. New York: St. Martin's Press, 1988.

————. "Bananas, Bases, and Patriarchy." In *Women, Militarism, and War: Essays in History, Politics, and Social Theory,* edited by Jean Bethke Elshtain and Sheila Tobias. New York: Rowman & Littlefield, 1990.

————. *The Morning After: Sexual Politics at the End of the Cold War.* Berkeley: University of California Press, 1993.

Evans, Sara M., and Harry C. Boyte. *Free Spaces: The Sources of Democratic Change in America.* Chicago: University of Chicago Press, 1992.

Faludi, Susan. "The Naked Citadel." *New Yorker,* 5 September 1994, 62–81.

Faulkner v. Jones. 10 F.3d 226 (4th Cir. 1993).

Flynn, Kevin, and Gary Gerhardt. *The Silent Brotherhood: The Chilling Inside Story of America's Violent Anti-Government Militia Movement.* New York: Penguin Books, 1995.

Franklin, John Hope. *The Militant South.* Cambridge: Belknap Press of Harvard University Press, 1956.

Garber, Marjorie. *Vested Interests: Cross-Dressing and Cultural Anxiety.* New York: HarperCollins, 1992.

Geerken, John H. "Machiavelli Studies Since 1969." *Journal of History of Ideas* 37 (1976): 351–68.

Gilbert, Allan H. *Machiavelli's* Prince *and Its Forerunners.* Durham, NC: Duke University Press, 1938.

Gilligan, Carol. *In a Different Voice: Psychological Theory and Women's Development.* Cambridge: Harvard University Press, 1982.

Godineau, Dominique. "Masculine and Feminine Political Practice during the French Revolution, 1793-Year III." In *Women and Politics in the Age of the Democratic Revolution,* edited by Harriet B. Applewhite and Darline Levy. Ann Arbor: University of Michigan Press, 1993.

Grazia, Sebastian de. *Machiavelli in Hell.* New York: Vintage Books, 1989.

Hartsock, Nancy. *Money, Sex, and Power: Toward a Feminist Historical Materialism.* Boston: Northeastern University Press, 1983.

Hartz, Louis. *The Liberal Tradition in America.* San Diego: Harcourt Brace Jovanovich, 1955.

Hegel, G. W. F. *Hegel's Political Writings.* Translated by T. M. Knox. Oxford: Clarendon Press, 1964.

Honig, Bonnie. *Political Theory and the Displacement of Politics.* Ithaca: Cornell University Press, 1993.

Hulliung, Mark. *Citizen Machiavelli.* Princeton: Princeton University Press, 1983.

Huntington, Samuel P. *The Soldier and the State: The Theory and Politics of Civil-Military Relations.* Cambridge: Belknap Press of Harvard University Press, 1957.

James, William. "The Moral Equivalent of War." In *Education for Democracy,* edited by Benjamin R. Barber and Richard Battistoni. Dubuque, IO: Kendall/Hunt, 1993.

Jeffords, Susan. *The Remasculinization of America: Gender and the Vietnam War.* Bloomington: Indiana University Press, 1989.

Kaminski, John P., and Richard Leffler, eds. *Federalists and Antifederalists: The Debate Over the Ratification of the Constitution.* Madison, WI: Madison House, 1989.

Karl, Jonathan. *The Right to Bear Arms: The Rise of America's New Militias.* New York: Harper Paperbacks, 1995.

Landes, Joan. *Women and the Public Sphere in the Age of the French Revolution.* Ithaca: Cornell University Press, 1988.

Lerner, Max. *The Prince and the Discourses.* New York: Random House, 1950.

Levy, Darline, and Harriet Branson Applewhite. "Women and Militant Citizenship in Revolutionary Paris." In *Rebel Daughters: Women and the French Revolution,* edited by Sara E. Mezler and Leslie W. Rabine. New York: Oxford University Press, 1992.

———. "Women, Radicalization and the Fall of the French Monarchy." In *Women and Politics in the Age of the Democratic Revolution,* edited by Harriet B. Applewhite and Darline Levy. Ann Arbor: University of Michigan Press, 1993.

Levy, Darline; Harriet Branson Applewhite; and Mary Durham Johnson; eds. and trans. *Women in Revolutionary Paris 1789–1795: Selected Documents.* Urbana: University of Illinois Press, 1979.

Machiavelli, Niccolo. *Machiavelli: The Chief Works and Others.* Translated by A. Gilbert. Durham, NC: Duke University Press, 1965.

Mahon, John K. *History of the Militia and the National Guard.* New York: Macmillan, 1983.

Mansfield, Harvey C., Jr. "Strauss's Machiavelli." *Political Theory* 3 (1975): 372–84.
———. *Machiavelli's Virtue*. Chicago: University of Chicago Press, 1996.
Martin, Alfred von. *Sociology of the Renaissance*. New York: Harper & Row, 1963.
McGerr, Michael E. *The Decline of Popular Politics: The American North, 1865–1928*. New York: Oxford University Press, 1986.
Meinecke, Friedrich. *Machiavellism: The Doctrine of Raison d'État and Its Place in Modern History*. Translated by Douglas Scott. New Haven: Yale University Press, 1957.
Millett, Allan R., and Peter Maslowski. *For the Common Defense: A Military History of the United States of America*. New York: Free Press, 1984.
Moser, Richard R. *The New Winter Soldiers: GI and Veteran Dissent during the Vietnam Era*. New Brunswick, NJ: Rutgers University Press, 1996.
Moskos, Charles C. *A Call to Civic Service: National Service for Country and Community*. New York: Free Press, 1988.
Moskos, Charles C., and John Sibley Butler. *All That We Can Be: Black Leadership and Racial Integration the Army Way*. New York: Basic Books, 1996.
National Alliance. "The Saga of White Will." *New World Order Comix*. no. 1. Hillsboro, WV: National Vanguard Books, 1993.
———. "National Alliance Goals." National Alliance Main Page, http://www.natvan.com.
———. "What Is the National Alliance?" National Alliance Main Page, http://www.natvan.com.
National Vanguard Books staff. *Who Rules America?* Hillsboro, WV: National Vanguard Books, 1993.
Nolte, Ernst. *The Three Faces of Fascism*. Translated by Leila Vennewitz. New York: Holt, Rinehart and Winston, 1966.
Nussbaum, Martha. "The Professor of Parody: The Hip Defeatism of Judith Butler." *New Republic,* 22 February 1999.
Ostrovsky, Victor, and Claire Hoy. "By Way of Deception Thou Shalt Do War," *Race and Reason* 1 (Jan./Feb. 1993). Posted as of May 1999 on http://www.stormfront.org. Originally published in *National Vanguard Magazine*, P.O. Box 330, Hillsboro, WV 24946.
Ozouf, Mona. *Festivals and the French Revolution*. Translated by Alan Sheridan. Cambridge: Harvard University Press, 1988.
Palmer, R. R. *The Age of the Democratic Revolution*. 2 vols. Princeton: Princeton University Press, 1959, 1964.
Pitkin, Hannah Fenichel. *Fortune Is a Woman: Gender and Politics in the Thought of Niccolo Machiavelli*. Berkeley: University of California Press, 1984.
Pocock, J. G. A. "A Comment on Mansfield's 'Strauss's Machiavelli.' " *Political Theory* 3 (1975), 385–401.
———. *The Machiavellian Moment: Florentine Political Thought and the Atlantic Republican Tradition*. Princeton: Princeton University Press, 1975.
Pole, J. R. *The American Constitution For and Against: The Federalist and Anti-Federalist Papers*. New York: Hill and Wang, 1987.
Reardon, Betty. *Sexism and the War System*. New York: Teachers College Press, 1985.
Reich, Wilhelm. *Mass Psychology of Fascism*. New York: Farrar, Straus and Giroux, 1970.
Rousseau, Jean-Jacques. *Politics and the Arts: Letter to M. d'Alembert on the Theatre*. Translated by Allan Bloom. Ithaca: Cornell University Press, 1960.

————. *Discourse on the Sciences and Arts (First Discourse)*. Edited by Roger D. Masters and translated by Roger D. and Judith R. Masters. New York: St. Martin's Press, 1964.

————. *La Nouvelle Heloise: Julie, or the New Eloise*. Translated and abridged by Judith H. McDowell. University Park: Penn State Press, 1968.

————. *On the Social Contract*. Edited by Roger D. Masters and translated by Judith R. Masters. New York: St. Martin's Press, 1978.

————. *Emile; or, On Education*. Edited by Allan Bloom. New York: Basic Books, 1979.

————. *The Government of Poland*. Translated by Willmoore Kendall. Indianapolis: Hackett, 1985.

Sandel, Michael J. *Democracy's Discontent: America in Search of a Public Philosophy*. Cambridge: Belknap Press of Harvard University Press, 1996.

Schudson, Michael. "Was There Ever a Public Sphere? If So, When? Reflections on the American Case." In *Habermas and the Public Sphere,* edited by Craig Calhoun. Cambridge: MIT Press, 1992.

Shalhope, Robert E. "The Armed Citizen in the Early Republic." *Law and Contemporary Problems* 49 (Fall 1986): 127–41.

Shapiro, Joseph P. "An Epidemic of Fear and Loathing: Bar Codes, Black Helicopters and Martial Law." *U.S. News and World Report,* 8 May 1995.

Shklar, Judith. *Men & Citizens: A Study of Rousseau's Social Theory*. Cambridge: Cambridge University Press, 1969.

Skinner, Quentin. *Machiavelli*. Oxford: Oxford University Press, 1981.

Skowronek, Stephen. *Building a New American State: The Expansion of National Administrative Capacities, 1877–1920*. Cambridge: Cambridge University Press, 1982.

Smolowe, Jill. "Enemies of the State." *Time,* 8 May 1995.

Snyder, R. Claire. "Shutting the Public Out of Politics: Civic Republicanism, Professional Politics, and the Eclipse of Civil Society." *An Occasional Paper of the Kettering Foundation*. Dayton, OH: The Charles F. Kettering Foundation, 1999.

Steger, Manfred. "The 'New Austria,' the 'New Europe,' and the New Nationalism." Paper presented at the annual meeting of the American Political Science Association, Washington, DC, September 1993.

Steger, Manfred, and F. Peter Wagner. "Political Asylum, Immigration, and Citizenship in the Federal Republic of Germany." *New Political Science* 24/25 (Summer 1993): 59–73.

Sternhell, Zeev, with Mario Sznajder and Maia Asheri. *The Birth of Fascist Ideology: From Cultural Rebellion to Political Revolution*. Translated by David Maisel. Princeton: Princeton University Press, 1994.

Stiehm, Judith Hicks. *Arms and the Enlisted Woman*. Philadelphia: Temple University Press, 1989.

Strauss, Leo. *Thoughts on Machiavelli*. Glencoe, IL: Free Press, 1958.

Talmon, J. L. *The Origins of Totalitarian Democracy*. London: Secker & Warburg, 1952.

Theweleit, Klaus. *Male Fantasies*. 2 vols. Translated by Stephen Conway in collaboration with Erica Carter and Chris Turner. Minneapolis: University of Minnesota Press, 1987.

Tobias, Sheila. "Shifting Heroisms: The Uses of Military Service in Politics." In *Women, Militarism, and War: Essays in History, Politics, and Social Theory,* edited by Jean Bethke Elshtain and Sheila Tobias. New York: Rowman & Littlefield, 1990.

Tocqueville, Alexis de. *Democracy in America.* New York: New American Library, 1956.

Trelease, Allen W. *White Terror: The Ku Klux Klan Conspiracy and Southern Reconstruction.* Baton Rouge: Louisiana State University Press, 1971.

United States v. Commonwealth of Virginia, 44 F.3d 1229 (4th Cir. 1995).

United States v. Commonwealth of Virginia, 976 F.2d 890 (4th Cir. 1992).

United States v. Commonwealth of Virginia 766 F.Supp. 1407 (W.D.Va 1991).

Viroli, Maurizio. "The Meaning of Patriotism." Paper presented at the Walt Whitman Seminar, Rutgers University, New Brunswick, NJ, 1 February 1994.

Walzer, Michael. *What It Means to Be an American: Essays on the American Experience.* New York: Marsilio Publishers, 1996.

Weigley, Russell F. *History of the United States Army.* Bloomington: Indiana University Press, 1984.

Wheelwright, Julie. *Amazons and Military Maids: Women Who Dressed as Men in Pursuit of Life, Liberty and Happiness.* London: Pandora Press, 1989.

Wingrove, Elizabeth. "Sexual Performance as Political Performance." *Political Theory* 23, no. 4 (November 1995): 585–616.

Wood, Gordon S. *The Creation of the American Republic, 1776–1787.* Chapel Hill: University of North Carolina Press, 1969.

Young, Iris Marion. "Impartiality and the Civic Public." In *Feminism as Critique,* edited by Seyla Benhabib and Drucilla Cornell. Minneapolis: University of Minnesota Press, 1988.

Zeeland, Steven. *Sailors and Sexuality: Crossing the Line Between "Straight" and "Gay" in the U.S. Navy.* New York: Haworth Press, 1995.

Zerilli, Linda. *Signifying Woman: Culture and Chaos in Rousseau, Burke, and Mill.* Ithaca: Cornell University Press, 1994.

Index

virtu, 34, 37; connection to civic virtue, 10, 16, 18, 36; definition, 19; production of, 22–27, 33
virtue: Christian virtues, 25; martial virtues, 8, 23, 26, 149; necessary for self-government, 8; relationship of civic virtue to martial virtue, 24, 54, 137; and sexual desire in Rousseau, 63–64; connection to vices, 6, 8, 9, 143, 157, 159. *See also* vices
voters, 91
voting, 89

war, 8, 20, 21, 34–36, 46, 92–93, 146, 160. *See also* Vietnam
warrior culture, 7, 156
West Point, 82, 102n22, 147, 149
white rifle companies, 94–95
white supremacy, 93–95, 107, 109–13, 109, 117, 128, 130. *See also* Ku Klux Klan
Wingrove, Elizabeth, 56, 63
women, 2, 12, 25–26, 30–31, 50, 57; as citizen-soldiers, 12, 46, 69–72, 143; danger of *armed masculinity* for, 156; exclusion of, 31, 57–59, 62, 65, 138,

153; and the far Right, 110, 114, 129; possibility of republican citizenship for, 64, 137, 142–43, 165; and the public sphere, 142, 163–64; rationale for exclusion from military, 1, 150, 155; role in French Revolution, 69–72, 77n115; as soldiers, 7, 144, 147, 148, 156; "women of the people," 69–72. *See also* "fantasy of Woman," feminism, republican motherhood, and subversive transgender performances
working people, 93, 97–99; and the far Right, 117, 119–20

xenophobia, 11, 12, 95–96, 110

Young, Iris Marion, 49–51, 60, 73n10, 75n71, 162–64

Zeeland, Stephen, 152–54
Zerilli, Linda, 56–59, 62–63, 125
Zionist Occupation Government, 118, 123
ZOG. *See* Zionist Occupation Government

About the Author

R. Claire Snyder is assistant professor of political science at Illinois State University, where she teaches courses in political theory and in women and politics. Currently, her primary areas of specialization are democratic and feminist theory. In addition to *Citizen-Soldiers and Manly Warriors,* Snyder has also published two articles, "Shutting the Public Out of Politics: Civic Republicanism, Professional Politics, and the Eclipse of Civil Society" and "The Public and Its Colleges: Reflections on the History of Higher Education." Her current research projects include continued work on the relationship between higher education and public life, as well as a new project examining the historical role of religion in America's progressively multicultural civil society.

Snyder is an associate of the Kettering Foundation and the former director of its projects on the Evolution of Public Life in American Democracy and the History of Higher Education. She has also worked extensively with the Walt Whitman Center for the Culture and Politics of Democracy at Rutgers University, where she directed the service learning project, Art Matters Too! and coordinated the Whitman Seminar. Snyder received her Ph.D. from Rutgers University in 1997 and her B.A. cum laude from Smith College in 1986.